The Thermal Warriors

BERND HEINRICH

The Thermal Warriors

Strategies of Insect Survival

HARVARD UNIVERSITY PRESS

Cambridge, Massachusetts, and London, England

First Harvard University Press paperback edition, 1999

Library of Congress Cataloging-in-Publication Data

Heinrich, Bernd, 1940–
    The thermal warriors : strategies of insect survival
/ Bernd Heinrich.
        p.   cm.
Includes bibliographical references (p.   ) and index.
ISBN 0-674-88340-3 (cloth)
ISBN 0-674-88341-1 (pbk.)
1. Insects—Physiology.   2. Body temperature—Regulation.
3. Insects—behavior.   I. Title.
QL495.H39   1996
595.7´01—dc20   96-1565

Designed by Gwen Frankfeldt

Contents

Preface

We believe that it requires great enthusiasms to deal accurately with little things; and that it is, consequently, impossible to meet with a reasonable or sober entomologist.

—*Edinburgh Review,* 1822

On a spring day in the mid-1970s, the sports page of the Boston *Globe* ran a picture of a runner in full stride. The runner was Jack Fultz, on his way to victory in the Boston Marathon, and he was squirting a stream of water from a plastic bottle onto his head as he ran. I doubt the *Globe* was trying to make a point, but in fact there might be a connection between the runner's cooling-off strategy and his victory. It had been an "exceptionally" hot day, and in a race with hundreds of nearly equally superb athletes, any one of whom could win, the margin between first and second place can be a matter of seconds. Over 26.2 miles that margin is determined by trifles, but temperature regulation is not a trifle. Athletic performance—indeed, all bodily activity—is the result of the interplay of vastly complex physiological processes, and all of them depend on temperature. All of the Boston Marathoners warmed up for maximal performance before the race started. Some ended up not at the finish line but in an ambulance, because of heat prostration, and many others trailed Fultz by minutes or hours. Against a field of supremely conditioned athletes, Fultz may have had an edge in temperature regulation.

When in the summer of 1983 I ran the final of 625.6 total loops on Bowdoin College's quarter-mile track to set an American record in the 24-hour run, I remembered the winning photo of Fultz. My "marathon" also took place on a hot day, and I dunked my head in a barrel of water alongside the track at every other lap.

We humans engage in endurance contests such as these only rarely, for the sheer fun or foolishness of it or for some symbolic trifle. For insects, however, the struggle to keep body temperature within an acceptable range is constant, and often it is a matter of life or death. Each insect is a "thermal warrior" in a contest with its predators and competitors in the context of its physical environment.

Consider the thermophilic ("heat-loving") *Cataglyphis* ants of the Sahara Desert that Rüdiger Wehner and his associates at the University of Zürich have studied. These ants, they found, are specialist foragers for heat-killed insects. The ants don't leave their subterranean colonies until it is so hot that they can be assured of finding dead and dying prey. But even these potentially lethal temperatures are not high enough—an evolutionary "arms race" has pushed the limits of safety. The ants wait still longer until even their own predators, heat-loving lizards, have been forced to retreat because of the heat. Only then, when temperatures are near the daily high point, which is near their own death point, does the colony march forth *en masse* to scour the searing sands for dead victims.

Masato Ono in Japan reports on another thermal warrior, the Japanese honeybee *(Apis cerana japonica)*. The honeybees are heavily preyed on by the thickly armored giant hornets *(Vespa mandarina japonica)*. Hovering near hive entrances, these hornets snatch bees, chew off their wings and legs, and then macerate them to a pulp to feed to their larvae. The bees' stings are useless against the hornets, which have evolved heavy armor, but the bees have evolved a potent counter-strategy against the armored killer wasps. They post a large number of guards at their nest entrances, and if a hornet ventures too close, the guards rush out, clasp onto the hornet with their mandibles, and in seconds form a ball of bees around the hornet. Now the bees begin to shiver and the temperature inside the ball quickly rises to near 50°C. The hapless hornet struggles fiercely, unavoidably generating more heat of its own that is now trapped because of the bees, and it is soon dead from heat prostration.

Direct confrontations are not the only thermal battles in the insect world. Here in the New England forests, some moths have evolved thick insulation and the amazing capacity to shiver and keep warm in the cold, when their predators have migrated south or are in hibernation. And in the hot southwestern deserts, as Eric C. Toolson and associates have discovered, a sweating response allows the desert or "Apache" cicada to sing at noon on the hottest days, in the hottest part of the year, and in the hottest part of its range, while its vertebrate predators are unable to tolerate the heat and so cannot hunt it.

Temperature is extremely important to the everyday activities of almost all insects. A bumblebee, for example, that rests at a body temperature of 5°C must increase her metabolic rate 1,500 times in order to fly, but this higher metabolic rate is possible only after she first raises her thoracic temperature to near 40°C. Each insect can survive only at its own specific temperature or range of temperatures. And in each case there is meaning behind the particular requirements for body temperature.

As has been long explored by many vertebrate biologists, especially most recently by George A. Bartholomew and William A. Colder III, a major consideration in any discussion of body temperature and metabolic rate is body mass. This is especially pertinent in insects. From our perspective as vertebrates, one of the more obvious features about insects is their small body size, although among insects that size range is about as large as the range from shrew to elephant. Other things being equal, large animal bodies cool more slowly than small ones. For example, a person-sized body with a temperature of 40°C at an ambient temperature of 10°C cools (if it has stopped producing heat) at only some small fraction of 1°C per minute. In contrast, a 100-mg bumblebee under the same situation cools at about 1°C *per second,* or a rate at least 100 times greater. The larger body would not approach the same rapid cooling rate as the bee's even at absolute zero (−273°C), the lowest temperature in the universe. Maintaining a stable body temperature, as we and many live animals do, is essential for most living beings, and some of the larger-sized insects that do so accomplish a considerably greater feat than we humans—in whom a high body temperature is in part an inevitable consequence of large size—can claim for ourselves.

In a small mass, where every point of the body is close to the exterior, the heat loss from the interior is very rapid. By the same principle, however, a smaller body also *gains* heat much faster than a large one when the heat source is external. Thus, a small insect can *heat up* several degrees per minute when perching in a sunfleck, an option that is not available to us. Size is of paramount importance to insects because almost everything in their lives depend on it. Since insects vary over 500-thousand-fold in mass, and since they can vary their metabolic rate over a thousand times, depending on flight activity and temperatures, we can expect to find vastly different kinds of thermal strategies in the insect world.

I have studied insects from the standpoint of a physiologist and an ecologist, and I have marveled at their sophistication from the broader perspective of an evolutionary biologist. I have been rewarded by the thrills of discovering several novel physiological mechanisms insects use to maintain an appropriate body temperature and to enhance their endurance in flight. Indeed, one of the mechanisms of honeybees looks very much like Jack Fultz's own strategy. I count among my greatest joys the satisfaction of discovering, or seeing others discover, the ingenuity for survival that insects have evolved in the irreducible crucible of temperature. This is entertainment of the highest sort.

Insects are marvels of creation with a body plan utterly different from ours. If it were not for them and their arthropod kin (crabs, spiders, scorpions), could we even imagine that it is possible to wear one's skeleton externally, cast it off occasionally, and exchange it for a new one? Could we ever have dreamed of organisms changing completely from a worm-like form eating leaves to a brilliantly colored flying marvel, or metamorphosing from a squat, creeping, aquatic troglodyte to the world's most versatile flyer able to snatch mosquitos out of the air? Could we conceive of being born ready to respond "perfectly" to amazing details of one's environment, without having to spend a lifetime slowly acquiring the appropriate responses by learning? Contemplating these incredibly diverse gems of the natural world, one is impressed with the realization that, seen objectively, insects are perhaps the most, or one of the most, highly evolved form of life. Given the hundreds of millions of years that they have been in existence on this earth, and the very short time, often measured in weeks, not decades, they need to produce a new generation, it is plain to see that insects have had the opportunity to evolve considerably more than we have. They evolved flight, building architecture, and complex social systems probably hundreds of millions of years before any other organisms on earth had done so. Their amazing diversity and their perfection of design seem to be ample evidence for their high degree of evolutionary success. If any animals have explored the diversity of ways of coping with temperature, it is surely them.

The insect world is an exemplar of Life that has evolved on a track different from or at a wide tangent to our own vertebrate line. Wherever Life exists, however, it develops according to specific universal principles. All matter obeys the same physical laws, and one might learn much about mechanical devices, or vice versa, by studying nature's design in the insects. Tucked into hundreds of obscure and technical journals is a store of knowledge that represents the work of explorers who have seen and described new territory. But there is still much more to see. The far reaches of the Colorado River and the Amazon, and even the topography of Mars and the moon, have been explored, but many of the peaks and canyons of the insects' physiology and ecology are hardly mapped out.

In a previous book, *The Hot-Blooded Insects,* I devoted nearly 600 pages to synthesizing and reviewing a large field of study from several perspectives. In each chapter I examined one of the many different kinds of insects—from giant moths to tiny fleas. I felt that if I had organized it according to general principles, then many of the fascinating details and group-specific unique features would be neglected or lost. It was also necessary to trace the confusing cross-currents in the specialist literature. That task has been done.

My attempt here is to write a short book that is necessarily sketchy and incomplete. It summarizes the main points of the previous book and incorporates a number of new developments. It was inspired by a book of only 109 pages that has had a great influence on me and that I still cherish. It was the Christmas present from my father when I was sixteen years old, and it was inscribed "Bernd Heinrich, dem Imker von seinem Vater zu Weihnachten 1956."

As the inscription indicates, my father duly acknowledged me as an "Imker" (beekeeper) even though I had only a hive or two. I did, however, spend as much time as possible on late summer days in western Maine learning the art of "lining" bees to find bee trees with my friends, the Adamses and the Potters.

Karl von Frisch's little book, titled *Bees, Their Vision, Chemical Senses, and Language,* could not have been more apropos. I even read it. Had I instead received his 566-page book on the same topic, I'm afraid that the wonder of it all might have escaped me. Certainly the book would not have been read. Indeed, I regretfully admit that my copy of the larger tome sits prominently on my bookshelf but is still unread.

It was von Frisch's *little* book that raised my eyes from the goldenrods and the pollen-laden bees flying off to old hollow hemlock trees in the woods. It was the little book that made me aware of new vistas. By way of simple and elegant experiments, von Frisch described the sensory world of these creatures and their ways of communicating in a surprising symbolic language. A new world was revealed to me, one that would otherwise have been blocked from my consciousness. The big book is a solid and valuable reference and a reminder of what stands behind the wonders that exist in the world. But nothing that I have ever read about bees has ever eclipsed the surprise and joy of learning how the author discovered what the bees sense and how they communicate with hivemates. As Donald R. Griffin stated in the introduction to the book, "The thought and the word are closely linked together; and for this reason an effort has been made to preserve in the printed page something of the pleasing directness and simplicity so apparent in the original lectures" (on which the book is based).

In later years I have been all the more mesmerized by the straightforward simplicity and purity of the little book, and in no small part I have felt its genius was due to the author's ability to leave out everything but the bare essentials. Beauty is the uncompromising economy that seeks only to illuminate truth, without obeisance to anyone, and without acknowledgment to stray thoughts or distracting concerns.

Thermoregulation perhaps no longer lends itself to uncompromising focus. On the other hand, brevity itself is a virtue to strive for, and it may be especially so in these more hurried times. I have therefore tried to concentrate on telling examples that illustrate general principles for readers who may not previously have been exposed to the field. Some aspects of insect thermoregulation may seem esoteric, but there is much of general relevance to be found by studying insects, and the study of body-temperature regulation during flight offers analogies with familiar principles of motor mechanics and vehicle maintenance. As Robert Pirsig said in *Zen and the Art of Motorcycle Maintenance,* beauty is not in what is seen but in what it means. You cannot "take apart" an insect that is smaller than your fingertip the way you can take apart a motorcycle. That has made the discovery of the marvels of design responsible for insects' extraordinarily high physical performance all the more fascinating.

In this primer I try to explain how, when, and in general what insects regulate body temperature. I also give an evolutionary perspective on the meaning of thermoregulation: why some insects (but not others) came to regulate high body temperatures. Morphology is the handmaiden of physiology, and it is essential to examine relevant insect flight-motor construction before describing specific passive and active heating and cooling mechanisms. All this is background to a discussion of the strategies of the thermal warriors themselves. To help the reader continue exploring an interest I hope this book will spark, I provide a selected list of references to the original scientific literature on the topics discussed. And to emphasize the "whole animal" approach (one that addresses physiological questions in both the ecological and evolutionary context), I have included pencil sketches of my favorite subjects.

Numerous people helped to shape this book. Michael Fisher, of Harvard University Press, encouraged me to begin writing it, and suggestions made by Doekele G. Stavenga and by two anonymous reviewers helped me revise earlier drafts. Their efforts are greatly appreciated. Erika Geiger and Suzanne Markloff were able to decipher my handwriting and patiently typed several versions.

I profited greatly from discussions with Brian Barnes, Harald Esch, Martin E. Feder, Joe Tallon, Jr., Hayo H. W. Velthius, Adriaan van Doorn, Robert D. Stevenson, Brent Ybarrondo, and James E. Marden. Their comments and especially the unpublished materials they made available were invaluable. Other unpublished information that was also useful was generously provided by R. L. Anderson, Madeline Beekman, Lincoln Brower, Joseph R. Coelho, John R. B. Lighton, Allen J. Ross, W. A. Woods Jr., Annemarie M. Surlykke, and Asher E. Treat. Finally, I express gratitude to my editor, Kate Schmit, of Harvard University Press, who painstakingly went through the whole manuscript to smooth out the rough spots.

The Thermal Warriors

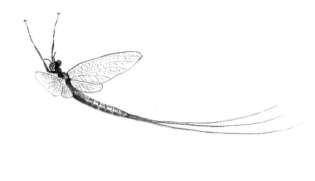

1

From Cold Crawlers to Hot Flyers

Insects arose on earth at least 350 million years ago, in the Devonian Period of the Paleozoic Era. Little is known about the earliest forms, except that originally they must have been crawlers, not flyers. And since all crawlers today, whether arthropods, mollusks, or annelids (earthworms and leeches), are "cold-blooded," it is likely that these first insects were too—that is, their bodies assumed approximately the temperature of the immediate surroundings. This holds true even when the immediate surroundings are quite frigid. The adult form of a midge (*Diamesa* sp.) walks on glacier ice even when its body temperature is chilled to −16°C. It is so sensitive to heat that, when taken from its natural environment and held in one's hand, it is killed by the warmth of one's skin.

Today, however, there are insects that maintain, or thermoregulate, a high body temperature. Some species of sphinx moths, for example, have thick insulating fur and normally maintain a thoracic temperature near 46°C over a wide range of ambient temperatures. (An insect's body temperature may be measured from the head, the thorax, or the abdomen; if the exact location is unimportant, I will occasionally refer to "body temperature," but in most cases the distinction is crucial and one must specify where the temperature was recorded.) Compared with these moths, our own normal body temperature of 37°C is almost cool. Most insects exist under conditions somewhere in between the cold-blooded crawler and the hot-blooded flyer, but these extremes show us what is possible, and they thus offer us a remarkable window into thermal adaptation from an evolutionary perspective. What we learn from them may have relevance for animals ranging from dinosaurs to dugongs.

Our earliest perspectives on the meaning of having an elevated body temperature came from observing ourselves and our warm-blooded vertebrate relatives. The ability of birds and mammals to regulate body temperature at one setpoint, specifically 37–41°C, has long been considered proof of sophistication and phylogenetic advancement relative to animals whose body temperature varies with that of their environment. Deviation of body temperature from the set-point of 37–41°C is, in birds and mammals, often associated with illness, and was once thought to be due to failure of the thermoregulatory system. Of course we *do* heat up with a "fever" when we are sick and cool down in a torpor when we are near death. We now know, however, that both increases and decreases in body temperature can be and often are adaptive responses. Both responses are often sophisticated physiological mechanisms that involve *more* thermoregulation rather than less, albeit the body is kept at a more appropriate temperature for specific conditions.

Given a functional rather than a strictly mechanical definition of *thermoregulation* that takes cost-benefit ratios into account, one can discard the idea that animals "should" at all times and all external temperature conditions maintain a single, specific body temperature. One can reason instead that although many insects show no evidence of strict thermoregulation (never varying from the setpoint temperature), others have superb thermoregulatory control. This new perspective of evolutionary fine-tuning opens us to new generalizations and insights.

Evolutionarily, insects are notoriously flexible relative to us, because hundreds of insect generations are produced for each new human generation, and insects have a large reproductive output. For example, fruit flies are both prolific and rapid reproducers (about 700 generations in the span of one of ours). Even if 97 percent of all fruit flies die in response to a specific selective pressure (such as a pesticide or temperature stress), then the descendants quickly regain the previous population level, and most will be resistant to what once was a bottleneck to their survival.

Within each of the diverse insect groups—moths, flies, wasps, bees, beetles, and cicada and their relatives—there are some species that fly with a thoracic temperature as low as 0°C, the freezing point of water, while close relatives within the same groups that are large-bodied vigorous flyers operate at body temperatures near 45°C. Phylogenetic arguments alone, therefore, do not help explain the pattern of body temperature along a gradient of evolutionary advancement.

The insect's most commonly regulated body part is the thorax, where the wings and flight muscles are located. Whether or not an insect regulates its thoracic temperature, and how it does so or when, or how precisely and at what temperature, seems to depend solely on the animal's mass and flight activity—or, more precisely, on the relationship between its flight metabolism and its size. Thus, understanding insect thermoregulation requires not only an appreciation of its size but also an understanding of the evolution of insect flight. The evolution of insect thermoregulation is inseparable from both.

The problem of insect-flight evolution is complicated, however, by the fact that the wings have become highly modified in many groups and now serve all sorts of other functions. These include sexual signaling by color markings (some dragonflies, grasshoppers, butterflies), sexual signaling by sound patterns (katydids, crickets, some grasshoppers), armor (beetles), mimicry and camouflage (innumerable groups), sailing in the air (very small insects), sailing on water (some stoneflies), flight control by gyroscope-like modifications of the hind "wings" (flies), shielding against solar radiation (principally beetles, butterflies), and convection baffles (some dragonflies and butterflies). Theoretically, any one of these other functions could have been the original function for which wings evolved, and from which flapping flight might then have been secondarily derived. Since evolution proceeds in those directions where there are the most benefits and the least costs, one function generally "piggy-backs" on another during the course of evolution, and the two are not always easily separable. The problem for us is: which function evolved first because it offered a benefit, and which followed because of the free ride?

Insects evolved from many-legged annelid-like crawlers, possibly like the *Peripatus,* which still lives among the damp leaves in the tropics of Africa, Asia, and Australia. This creature reminds one of an earthworm or a leech, but it has a pair of short legs on each of its many segments. Sometime in the Carboniferous Period, about 300 million years ago, some of the annelid-like crawlers, perhaps something like a *Peripatus,* took a series of fateful evolutionary steps. The forms with one pair of legs per segment evolved an armored covering that at first probably served primarily for defense, as it did in the earliest vertebrate animals then also evolving in the coal swamps.

It had previously been supposed that fishes evolved legs because walking across land to new water holes bestowed an advantage when the swamps began to dry up. More recently it has been suggested that an "arms race" in the swamps made vertebrates there develop huge teeth and armor. In other words, the presence of many predators may have been the primary selective pressure for soft-skinned amphibian-like creatures to leave the fray in the water and win the battle by avoiding it, setting off the tetrapod explosion of new forms on land. Perhaps a similar scenario could have nudged the insects to leave the water as well, to start the evolution of wings to conquer the air.

The evolution of external armor for defense prior to coming onto land was a pre-adaptation for two other functions. That is, the hard covering turned out to have uses beyond those for which it was first adapted. First, the armor allowed for the secure attachment of muscles and struts that promoted the possibility for rapid locomotion. Second, the covering facilitated waterproofing, which became essential or at least useful for the invasion of the land. One scenario holds that predation in the water was countered by escape to vegetation above or close to the water, where the organisms also encountered vast unexploited plant resources for food. Finally, the crowning achievement that consolidated and extended the rich ecological possibilities for land invasion was the evolution of flight. Flight probably led to the incredible flowering of insect diversity into the millions of species that we see today, and it necessitated thermoregulation for some of these forms.

Wings could not have appeared fully functional from the first. They must have originated from structures that originally served some other function. It has been proposed that the first precursors of wings served as solar heating panels or as signaling devices between the sexes, but both theories face numerous problems and strong inconsistencies. One argument against these theories is that of the hundreds of thousands of present-day insect species that do *not* fly, none uses wing-like structures as solar panels to heat themselves, and none uses wing-like structures for sexual signaling. It seems unlikely, therefore, that such structures would have existed in the past even *before* wings evolved. Wings were used for these functions apparently only *after* flight had evolved.

The one theory that seems to be consistent with the evidence of paleontology, neuroanatomy, behavior, and embryology, is the one originally proposed by Sir Vincent Wigglesworth, the generally acknowledged "father" of insect physiology. His idea was that the wings of insects evolved from *gills* of aquatic mayfly- or stonefly-like ancestors.

In almost any stream or pond today you can still see the larvae of mayflies, caddisflies, dragonflies, damselflies, and stoneflies. Some of the larvae of mayflies (Ephemeroptera), thought to most closely resemble the stock from which *all* insects evolved, possess paddle-like gills that help them extract oxygen from water when simple diffusion through the armored body will no longer suffice. The larval mayflies of today resemble the ancient mayfly-like forms that were without wings even when adult and that must have reproduced and dispersed without metamorphosis. In present-day species, of course, the adult is the form after the last molt, and it is specialized for dispersal and reproduction.

In stagnant water, which has less oxygen than running water, moving the gills (by muscles from within the body) aids gas exchange. More vigorous gill motion not only moves more water around the animal, but it can also propel the animal; in fact, some present-day fish (sharks, tuna) *must* propel themselves in order to breathe. Similarly, mayfly larvae swim and breathe by gill paddling, and there is in

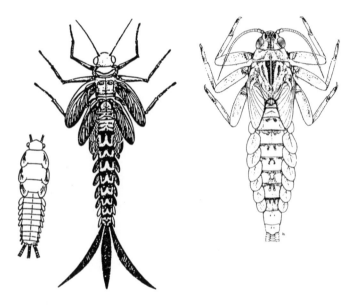

The two sketches at left depict an early and an older nymph of the late-Paleozoic mayfly-like insect, *Kukalova americana*. In these renderings you can see the development of lateral gills and of pre-wings that were possibly used in underwater locomotion. To the right is a nymph of a present-day mayfly (*Bactis* sp.), with tails deleted.

these insects only a small and arbitrary distinction between respiratory and locomotory functions of the gills. The paddle-like appearance of the gills as found in some mayfly species is therefore not incidental. What is less clear is what intermediate stages existed between rowing in water and rowing in air. Even to the present day, some species of stoneflies (Plecoptera) of the northeastern United States that emerge from partially ice-covered streams during the winter or early spring rarely if ever use their wings for flight. Instead, they use them for rowing on top of the water surface. Some of these stonefly species have very small wings that closely resemble the moveable thoracic gill lobes that were used both for ventilation and rowing during the Lower Permian, suggesting that rowing on top of the water may have been the critical step prior to flight.

It is also possible that the small wings of the winter-active stoneflies represent a reversion from fully flying ancestors to non-flying forms, as has occurred at numerous times in various winter-active flies and moths and in scorpion flies. Quite simply, gill-like structures could have served as crude wings, aiding in short flights after the insect first gained altitude by crawling up on aquatic vegetation. Even now we see insects that use and benefit from this type of locomotion. In these insects the flight apparatus is weak relative to the weight carried. The present-day examples of the adaptive use of short flights are seen in cicada killers, *Sphecius speciosus,* and spider wasps, *Pepsis* and *Ammophila,* insects from different families who carry immobilized prey long distances to their nests. To facilitate their work they crawl up grass stems (with prey) and then fly a little ways while rapidly losing altitude; by repeating the process they will eventually reach their destination. Early insects could have similarly benefited when they still had weak wings relative to the weight of their bodies, which they had to support in the air.

Prior to the evolution of long-distance flight, the larval-like animals resembling stonefly and/or mayfly nymphs never underwent metamorphosis but they still became reproductive and had to find mates and oviposition sites. There was no winged, "adult" dispersing stage

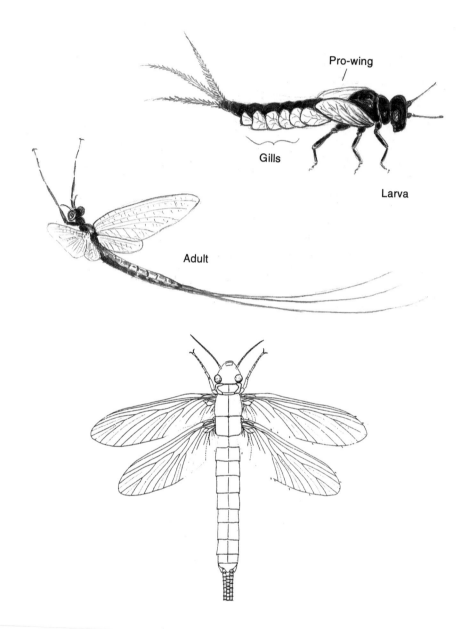

Above: Adult and larva of present-day mayflies (*Drunella* sp.). *Below:* Presumed adult of late-Paleozoic mayfly-like insect, *Kukulova americana.*

in their life cycle. But an ability to occupy new or uncontested habitat was undoubtedly as important then as it is now. Paddling ability was therefore likely *more* developed in these larval-like organisms than it is in present-day larvae.

Whatever the species, insect flight likely evolved only once, because the complete pattern of wing venation is similar in all the insect orders, suggesting that the pattern evolved from a single ancestor, namely the mayfly-like forms. The gills of the nymph- or larval-like ancient insects were similar in structure to insect wings today: two membranes folded together and enclosing hollow tubes or veins. But when hollow tubes become thickened to serve as struts for support (needed for strenuous flight), the gills because less well suited for gas exchange, the process by which the animals obtained oxygen. How, then, could two opposing functions (locomotion and respiration) be incorporated in one structure?

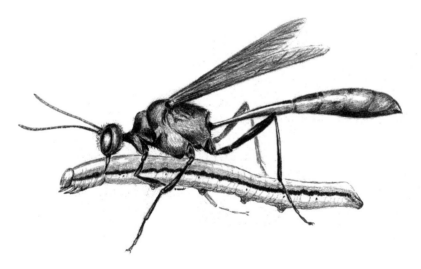

An *Ammophila* sp. wasp dragging a caterpillar to its burrow, where it will place an egg on the drugged, comatose prey. The grub that develops will devour the caterpillar. To transport large prey that can't be held aloft in level flight, the wasps often climb plants from which they can make short flights. Early insects with less developed wings possibly also made similar "jump-flights" to gain distance.

The solution in the ancient mayfly and stonefly nymphs was the evolution of separate structures. The leaf-like gills on the *thorax* became thicker, hence modified for better locomotion (presumably at the cost of reduced gas exchange), while the thoracic muscle mass to move them also increased. In general, there was an increase in size of the thoracic lobes and a decrease in number. In most mayflies there were only two pairs of wings, but in some species (Caenidae, Tricorythidae, Baitidae) even the second pair became small or absent as the first got larger. Meanwhile, the lateral paired gills or paddles on the *abdomen* remained thin enough to retain their primary respiratory function.

An increasing differentiation between a "thorax" anteriorly and an "abdomen" posteriorly evolved from the otherwise relatively uniform arrangement of numerous segments in worm-like ancestors. The head and thorax each resulted from the consolidation or fusion of three anterior segments. Posterior to the thorax the original segmentation was retained at least externally, as is evident in the overlapping body plates that we still see in all insect abdomens. Internally, these segments became consolidated into an area specialized for digestive and reproductive functions.

As the anterior respiratory air tubes became thickened for locomotion, the resulting decrease in oxygen uptake might have been compensated for both by increased "gill" size and movement and by migration out of the water into the oxygen-rich air (even richer then than now) just above the water. Indeed, stagnation of the water (which reduces its oxygen supply) and increasingly higher water temperatures (which increases the metabolic demands on the insects) might also have forced insects out of the water. Simultaneously, either short glides or short, violent bursts of powerful paddling on the surface of the water would have provided the triple advantage of facilitated dispersal, escape from predators, and contact with a vast untapped energy resource, terrestrial plants.

Although numerous selective pressures could have converged to alter the "gills," these selective pressures pulled in two directions at

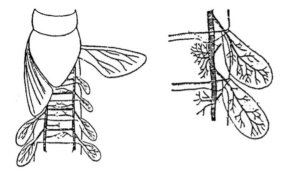

Gills and rudiments of thoracic wings in a present-day mayfly nymph. Note the resemblance of the gills to the pro-wings of the *Kukalova americana* nymph (illustrated above).

once. The land and water environments present radically different respiratory and locomotory demands on an organism, and it is difficult to imagine that one structure—a gill—could have two functions in just one environment, much less in two environments. But, in part because of their hard exoskeleton, insects had already evolved a mechanism—molting—that would yield a solution.

Insects and crustaceans must shed their exoskeletons in order to grow, and this regeneration of their outside covering and support system presents each individual with an opportunity to be "reborn." Molting allows an organism to grow to a slightly larger size but, more important, it also permits it to assume different body proportions: its paddles, for example, could *suddenly* become larger and more sturdy after molting. That is, from one molt to the next the anterior thoracic paddles could become wings while the posterior paddles could be lost entirely. (Over time, the three pairs on the thorax became reduced to two in most species, and the front pair, from the pro-thorax, was lost.)

It goes without saying that the evolution of flight of birds, bats, and pterosaurs followed an entirely different scenario. Nevertheless, the *initial* wings in vertebrates may first also have evolved independently from gliding structures, such as are now found in some fishes, rodents, marsupials, amphibians, and lizards. There is also precedence in vertebrates for the idea that limbs used for locomotion in water can be retooled for use in air, or vice versa. For example, the enlarged pectoral fins of flying fish now function as wings and allow the animals to leap out of the water to escape predators and to glide above the water's surface. Similarly, numerous species of birds (dippers, some ducks, guillemots, puffins, loons, auks, cormorants, shearwaters, and diving petrels) also use their wings both in air and in water and again prove that it is possible to use the same structure for locomotion in two media.

The two insect pre-adaptations—gill paddles and molting—likely touched off the insect "explosion" that is still going on today. In conjunction with leaving the water, insects obtained an advantage—flight—that aided their conquest of a vast array of new niches. All of them now faced the problem of a potentially wildly fluctuating body temperature.

In the competition for a place in the new environment, selective pressure would have resulted in explosive innovations. An analogy might be the development of the airplane: the first manned flight at Kitty Hawk quickly ushered in a parade of ingenious flying machines. Similarly, within only a few million years in the Permian Period, the ancient mayfly-like stock had radiated out to all the modern insect orders. Many species, however, still retain an aquatic larval stage that is relatively unchanged, although the functions of reproduction and dispersal were passed on to a new form, the adult, hundreds of millions of years ago. Examples include the stoneflies (from which moths and butterflies later evolved), dragonflies, damselflies, and caddisflies. But the invasion of the terrestrial habitat continued, until most larvae of the modern orders (flies, butterflies, wasps, and their kin) had escaped the water entirely.

By the time that the Carboniferous coal forests were being laid down, 300 million years ago, some insects' mastery of the air was complete: powerful flight had evolved. No more convincing evidence of this exists than the fossils of the giant dragonfly, *Meganeura,* which flew over the forests some 160 million years before the presumably clumsy attempts at flight by the birds' ancestor, *Archeopteryx.*

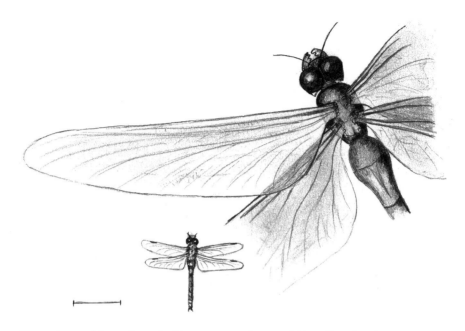

The extinct coal forest dragonfly, *Meganeura monyi*, about eighteen times heavier than one of today's largest dragonflies. (Scale = 5 cm.)

Meganeura existed in a variety of sizes, but the largest species, *M. monyi*, had a wingspread of close to a meter and likely weighed near 18 g, whereas today's largest dragonflies, such as the green darner *(Anax junius)*, only weigh up to 1 g. Furthermore, given its wing shape, mouthparts, and the placement of its legs, *M. monyi* was likely a swift flyer and an active hunter, like the present-day dragonflies. Dragonflies (and cockroaches) saw *Triceratops* and *Tyrannosaurus rex* come and go. These living fossils were so well adapted for flight soon after having evolved that one is forced to wonder whether improvements are still possible. Other insects, however, have taken divergent evolutionary paths. For example, many species no longer fly at all, having evolved the loss of wings to live in niches where flight is not needed or is too costly to maintain.

2

Heat Balance

Along with the "thermal war" that many insects have to fight against predators and prey, there is a need for them to maintain heat balance. They must balance the amount of heat they produce or absorb against the large amounts their small bodies lose to the environment in order to maintain a body temperature at which they can remain active in the competition for survival. Insects that *actively* adjust heat gain or loss are said to thermoregulate, and one of the most active steps they can take is the metabolic production of heat through muscle contraction.

In the same way that the motor heats up when a car burns fuel, heat is released as an inevitable by-product of cellular metabolism whenever muscle contracts. Most recent data show that close to 94 percent of the energy expended by muscles during contraction is degraded to heat, while about 6 percent appears as mechanical force on the wings. Insect flight is one of the energetically most demanding activities known, and thus most insects produce more heat per unit muscle mass when they fly than almost any organism on earth. Although there is no imprint of it in the fossil record, we can thus be about as certain as we are of their fossilized wing venation that some of the early insects were intermittently *endothermic*. That is, some of the large-bodied forms must have had a very high thoracic temperature due to the metabolism of the "flight motor," or wing muscles. This is not a matter of conjecture. It is instead a matter of biophysics.

Heat balance of a body (or a body part, such as a thorax) is described formally by the equation:

$$\frac{dH}{dt} = M - C\left(T_b - T_a\right)$$

Simply put, this equation expresses the change in heat in the body per unit time (dH/dt) as a function of heat produced and heat lost. The amount of heat produced (M, for heat input, such as metabolic heat) is measured in terms of the amount of oxygen the body uses for metabolism, which depends on body weight and time and so is expressed as so many milliliters of oxygen per gram of body weight per hour ($mlO_2/g/hr$). The amount of heat lost is calculated as a function of the body's conductance (C), a term that refers to the "leakiness" of heat, and the difference between the body's temperature (T_b) and air, or ambient, temperature (T_a). Conductance is affected by body size, wind speed, posture, insulation, blood circulation, and other factors. When body temperature remains constant (as when it is regulated at an elevated temperature), then $dH/dt = 0$, which is another way of saying that heat production equals heat loss, and $M = C(T_b - T_a)$. At this point $C = M/(T_b - T_a)$, so conductance can be expressed in terms of the metabolic demand ($mlO_2/g/hr$) required to maintain heat balance per degree centigrade that body temperature exceeds air temperature, or $mlO_2/g/hr/°C$. Metabolic rate (or heat input) can be calculated if the difference between body and ambient temperature ($T_b - T_a$, known as the temperature gradient) and the passive cooling rate (from which conductance can be calculated) are known. Conversely, from measurements of metabolic rate, one can calculate conductance and the difference between body and ambient temperature that animals of different size (and different conductance rates) can generate.

Small insects have much lower body temperatures in flight than large insects, not because they produce less heat—in fact, they may have *higher* rates of heat production than larger insects—but because they have much larger conductance due to their large relative surface area. In bees, for example, only the large species heat up in flight and generate an appreciable elevation of body temperature even though metabolic cost of flight per unit weight declines approximately 230 percent for a tenfold increase in mass. (Indeed, the weight economy with increasing size is why airlines try to ferry large numbers of passengers on one super-large airplane rather than on many small ones.) A mosquito in flight maintains only a tiny ($<1°C$) gradient between thoracic and ambient temperature, despite prodigious amounts of heat production. A blowfly *(Calliphora vicia)* may heat up 5°C, and a honeybee heats up its thorax about 15°C. Having a much larger thorax, and hence a smaller relative surface area, means that the internally generated heat during flight is not lost by convection at the same rate that it is produced until a much higher temperature gradient has been generated.

Over the last twenty years most internal temperature measurements of insects were taken with electronic probes on wire leads as thin as or thinner than a human hair. A recent innovation is thermovision, which gives a pictorial image that color-codes different body-surface temperatures. We now have a massive enough data set of body temperatures for insects varying in mass from less than 10 mg (smaller than a housefly) to more than 10 g (nearly mouse-sized) to see that in all groups the small species do not heat up appreciably. In those species whose body mass exceeds approximately 200–500 mg, the flight-motor temperature in full flight is often higher than our own body temperature. As already mentioned, some large sphinx moths even achieve a thoracic temperature in flight near 46°C, a body temperature that is not long tolerated by a human. The generation of these high temperatures is due both to the moths' large size (and therefore low conductance) and high metabolic rate (their small wings must beat very rapidly), aided in some species by a thick covering of insulating scales (which lowers conductance).

Muscle efficiency should increase rather than decrease by selection through evolution, hence the first insects that flew were probably not more efficient flyers than are those active today. They therefore produced as much as or *more* heat per given amount of mechanical work than present-day forms, and consequently they faced nearly the same thermal consequences of exercise as insects do now. That is, the larger they were and the more vigorously they flew, the hotter they became because of their lower conductance. But then as now there was likely a similar physiological limit on how high a body temperature they could tolerate.

Given its large size and extrapolating from data of present-day dragonflies, Michael L. May from Rutgers University has estimated that a *Meganeura monyi* in flight should have generated a thoracic-temperature excess (above air temperature) of over 62°C. Since this extinct giant dragonfly likely existed in an arid tropical environment, where air temperatures were close to 30°C, an excess this great would indicate (from the physical standpoint only) that the giant dragonflies reached a temperature of 92°C, within 8°C of the boiling point of water. Clearly they did not, however. Their bodies were constructed of proteins, as all animals today are, and they must have somehow regulated their body heat to a temperature nearly 50°C lower than 92°C. Even if we assume that they flew at the very unlikely low air temperature of 0°C, the 62°C generated by the flight motor would still have been unreasonably high for muscles as we know them.

Given what we now know, the large size of the extinct dragonfly, *Meganeura monyi,* becomes all the more remarkable. How could it possibly have reduced a potential temperature excess of over 62°C and cooled itself in flight? We do not know the behavior and physiology of this extinct creature. All we can do is extrapolate from modern large dragonflies. Modern large dragonflies—like the green darner, *Anax junius,* which weighs nearly one gram—probably provide a glimpse of what *M. monyi* could do or had to do to a much greater degree.

Anax junius is a consummate flying predator known all over North America. Like other dragonflies that heat up in flight, it shivers to warm up before taking off. It uses its abdomen as a "heat radiator" to get rid of excess heat when in flight at high ambient temperature; that is, it shunts heat to the abdomen, which then acts as a thermal window to increase heat loss. A possibility that cannot be excluded yet is that additional heat was lost as blood circulated through the wings, especially if they flew in shade. Modern dragonflies cannot use this option because, in their much smaller wings, the small wing veins do not bring much blood to the surface, where heat can be dissipated from it.

More recent work by May has provided convincing evidence that *A. junius* replaces continuous flapping with partially gliding flight under conditions of heat load. By doing so it reduces the heat load by about 30 percent. The extinct *M. monyi* had large wings, which are good for gliding, and could have used the same option.

We see many potential solutions in present-day insects. For example, in bees the heavily insulated species *(Bombus)* have invaded the Arctic and other cool areas. Large-bodied species that live in the tropics *(Euglossines)* are naked, and they fly at night when it is cooler, or they fly high and fast in the forest canopy, probably for rapid dissipation of heat. Physiological cooling mechanisms would also have been needed to keep body temperature within a reasonable range. What is more, since the large, extinct, strongly-flying insects were normally subjected to high thoracic temperatures in continuous flight, as I will show, they must also have been biochemically adapted to operate their flight muscles at the *high* temperatures experienced, and to have mechanisms for warming up the flight motor prior to flight. My aim in writing this book is to explore the insects' various mechanisms for temperature regulation, but it is first necessary to review some of the structures underlying these mechanisms.

3

The Flight Motor

In *The Nature of Nature*, the poet John Haines writes about the "fundamental relationship (of a sense of order or proportion that is right and hence beautiful)—in the way that an insect, for example, is put together, the way it functions. To anyone who has looked closely at a piece of machinery, the resemblance is obvious, and children often see this" (p. 199). The writer Robert Pirsig also champions the beauty of mechanical excellence, as seen by children, and sometimes also by entomologists. Indeed, the sense of wonder any of us feels when we encounter an insect is probably largely attributable to the perceived but unknown mysteries of how it is put together. To anyone not aware of these mysteries, the fly is only a bothersome irritation.

A fly is a marvelous miniature flying machine with a built-in pilot. It has two wings that row it through the air, a motor that powers the wings, a fuel tank, and a system of tubes and pumps to deliver the fuel. Other tubes and bellows deliver oxygen for the combustion of fuel in the flight motor and for the elimination of the carbon dioxide that is produced. It has chemo-, photo-, and mechanosensors and a nervous system that sends information via electrical signals to an integration center, which interprets the signals and then relays other signals to effectors that respond and control the whole system. It may also have temperature-control systems.

Since the thorax contains the powerhouse or "motor" that drives the insects' wings, it's not surprising that the thorax of most insects that heat up in flight is packed full of muscle. Teasing these muscles apart, you can see with the bare eye that they are compartmentalized into units. Half of the muscles (a set of two, a right and a left set) are in the center and run generally from the head toward the abdomen; these are the *dorsal longitudinal muscles,* or DLM, which, through a complex set of wing hinges, serve as the wing depressors. The other half of the muscles (also a right and a left

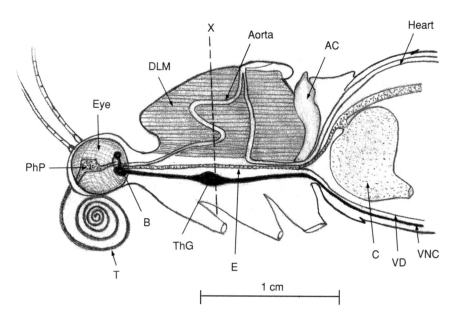

The muscles and organs of a sphinx moth (sagittal section). The dashed line labeled *X* indicates the position of the transverse section shown in the next illustration. *AC,* air chamber; *B,* brain; *C,* crop; *DLM,* dorsal longitudinal muscles (wing depressors); *E,* esophagus; *PhP,* pharyngeal pump; *T,* tongue; *ThG,* thoracic ganglion; *VD,* ventral diaphragm; *VNC,* ventral nerve cord.

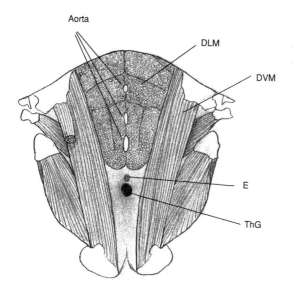

Aorta

DLM

DVM

E

ThG

Because it loops through the thorax, the aorta is exposed three times in this view *(transverse section)*. The dorso-ventral muscles *(DVM)* act as wing elevators.

set) run up and down, nearly at right angles to the DLM; these are the *dorso-ventral muscles,* or DVM, the wing elevators. (Dragonflies are an exception—both sets of muscles run up and down in these flyers.) The DLM and the DVM are the muscles that provide most of the power for the wing strokes. During flight (and pre-flight warm-up, discussed later), these two sets of muscles contract and relax alternately, operating analogously to two pistons or two sets of pistons in a two-cylinder gasoline engine. The wings do the next best thing to turning like a propeller. They subtly change their pitch during the wing stroke, so as to take a "bite" out of the air during the downstroke, as well as during the upstroke. The air intake and exhaust system (that is, the tracheal system) is separate from the fuel delivery system (the circulatory system, with hemolymph).

Muscles are able to harness large amounts of energy through structures that could be called the ultimate in miniaturization. In the gasoline engine the oxidation of fuel in a cylinder releases gases in a small explosion, and the gases then push a piston. In muscles, such as those of the insect flight motor, however, the gases (carbon dioxide and water vapor) contribute nothing to power. Instead, the energy released from the carbon-carbon bonds of the fuel is captured by a series of chemical transformations along millions of chains of flexible protein molecules arranged in series within hundreds of thousands of microscopic power-generating organelles, the mitochondria. The energy is harnessed into chemical bonds that the muscles translate into the mechanical force of contractions. The muscles accomplish work by repetitive *pulls*, rather than pushes.

Efficiency

Efficiency can refer to many things, but at the most basic level it concerns the ratio of work done (such as the mechanical performance of lift and thrust by the wings) to the total energy expended. Energy expenditure can be measured directly, in terms of fuel consumption, or indirectly, in terms of the amounts of oxygen consumed or carbon dioxide or heat released.

Although the flight muscles of insects, like the muscles of other animals, are at most 5–9 percent efficient in converting energy to useful work, they nevertheless have to be fine-tuned to extract even *this* much work to operate the wings. It's a matter of timing. Both gasoline engines and insect flight motors are powered by alternate strokes of force. One uses the power of a push and the other that of a pull, but for both the harnessing of energy that is applied to wheel, propeller, or wing requires that the power strokes remain truly alternate; they must not cancel each other out, for in that case they would be working against each other, resulting in zero efficiency. That is why we periodically have our cars tuned up, to make sure that the timing of the pistons going up at any one time does not interfere with the pistons going down. Greater efficiency in timing converts to better fuel economy and is a big selling point in the design of engines. Other things being equal, models with poor mechanical efficiency don't survive in the marketplace, and car engines have improved in efficiency over the decades. In insect flight the selection process for greater efficiency in insect flight has been going on for at least 300 million years, and as a result we have no reason to suspect the overall efficiency we see now is likely to get much better soon.

In both systems, mechanical efficiency is first of all a function of the mechanical design itself, but it is then fine-tuned through "physiology." In the gasoline motor, for example, the upstroke and downstroke pistons are mechanically linked to a stiff arm, the crankshaft, that prevents them from moving at random, and a firing coordinator, the distributor, is designed so that ignition occurs at the proper time of the piston cycle. The insect flight motor has similar provisions that ensure that the mechanical movements of the upstroke and downstroke muscles alternate during flight and do not occur simultaneously or at random. Alternate contraction is fine-tuned in many kinds of insects (the "neurogenic" flyers) by appropriate timing of neural commands from the nervous system.

As in gasoline engines, mechanical design is also of primary importance for efficient functioning in insect flight motors, and it can help reduce potential overlap in contractions. The flight motor of all insects is designed so that contraction of the set of upstroke muscles *automatically* stretches the downstroke muscles, thus preparing them for their own contraction. Indeed, in some of the more recently derived insects, such as the flies, wasps, and beetles, a stimulus for contraction of a muscle is stretching. The stretch stimulus eliminates the necessity for constant and precise neural signaling, wing beat by wing beat, from a pacemaker in the central nervous system. This "myogenic" mode of flight control allows some insects to beat their wings hundreds to over a thousand times a second, and mechanical inefficiency is greatly reduced because the upstroke or downstroke muscles contract only after their opposites have already contracted. Furthermore, the stretching of muscles and ligaments, like the stretching of rubber bands, stores energy, and this energy is then released on the next wing beat.

Temperature

Because muscle contraction is powered by chemical reactions, and because the speed at which chemical reactions take place depends on temperature, the speed at which any one muscle contracts is a function of its temperature. We can move our fingers only slowly when our hands are chilled on a winter day, but we can snap up a fly with them on a hot summer afternoon (provided the fly itself is not too warm, and thus too fast for us). Animals that cannot regulate their temperature and that therefore have body temperatures that change according to ambient temperature seem to be at the mercy of the weather; they are said to be *poikilothermic* ("variable in temperature"). For example, a poikilothermic ant may barely crawl on a cold morning but run swiftly in warm sunshine. Running behavior requires a concert of dozens or hundreds of different muscles in each of which twitch duration is strictly temperature-dependent. In many instances twitch duration, or the rate of movement, is not of overriding importance beyond affecting the pace of the activity. It may be critical for flight, however.

For powered flight it is essential that a minimum, often high rate of work output be maintained. Anything less is wasted effort. In order to achieve minimum wing-beat frequencies for takeoff, the muscles must contract rapidly enough. Temperature is also important for mechanical efficiency: at low muscle temperature there is partial overlap in contractions of the up- and downstroke muscles, and the two sets of muscles then work against each other rather than working to move the wings. Not all muscles are alike, however. Depending on the species, they have been evolutionarily designed to operate over ranges of either low or high temperatures.

We use various kinds of metals for different kinds of mechanical motors, depending on how hot we expect them to run. But insects must do largely with organic compounds, principally proteins. Proteins and other biological compounds are held together in complex units that must, like miniature motors, be able to move in precise ways in order to function. They bind molecules together and break molecular bonds, releasing reaction products, and then again change shape to return to the original conformation. Each molecule must be flexible, but not too flexible. It must be rigid, but not too rigid. Slight changes in temperature alter the delicate bonding of proteins, changing their structure and hence their capacity to perform. Because they indeed bend with changes of temperature, it may appear that proteins are designed to be sensitive to slight temperature changes, but they are not. Instead, they are temperature-sensitive primarily because of their need to be flexible, and so their temperature sensitivity cannot be avoided. Nevertheless, given an infinite range of temperature, proteins and other organic molecules can be modified or designed to be more or less rigid at one *specific* range of temperatures. The problem is, *which* temperature range should evolution choose, and how wide should the range be?

The degree to which proteins in insect flight motors are specialized for different temperatures is still largely a matter of speculation. In different species of moths, for example, the temperature characteristics of several enzymes catalyzing the breakdown of fuel to extract energy from it are nearly identical, even though thoracic temperatures range from 0°C to over 40°C. Nevertheless, those moths normally flying with a low muscle temperature also generate more power at these lower temperatures than those flying at high muscle temperatures. Furthermore, the low-temperature moths are able to fly at low temperatures also because of morphological adaptations, such as reduced weight and increased wing size, that decrease the power needed to stay airborne.

Regardless of what temperature the smallest biochemical units of the flight mechanism may be adapted to, they still need to be flexible to operate. The narrow temperature tightrope on which the flight motor balances shifts within seconds of takeoff in those cases where the body temperature rises quickly. Should evolution design proteins for the low temperature of the flight motor when the animal is at rest or warming up in preparation for flight, or for the high temperature characteristic of full flight? The answer relates to maximization of power output.

Power Maximization

An insect that is specialized for life on the ground, such as an ant or a carabid beetle, has no need to maintain a *specific* minimal rate of power output. Like a car, it can run slow or fast without immediate major consequences. On the other hand, for an insect such as a fly, a bee, or a moth, almost all biologically meaningful activity is predicated on flight. And as with propeller planes that require a very high minimum output of power for takeoff and flight, flying insects cannot beat their wings more slowly than a certain minimum speed or allow power output to decline below a (high) threshold. Indeed, the very high energy expenditure required for powered flight in insects is near the highest observed in the animal kingdom. This high level of power output is achieved by packing as many power-producing units into the available "engine" space as possible. For maximum efficiency the engine should therefore operate at one optimal temperature.

As an alternative to maximizing power output, consider the possibility that an insect might maintain temperature-*independent* activity by having, for example, a series of several parallel systems, each one optimal for a different temperature. This is like having several, say four, engines in the same propeller plane each constructed of different temperature-sensitive alloys and each operating maximally over a different temperature range. The plane can then always maintain the same power output regardless of what the air temperature is. But then, as different engines are turned on and off as the temperature changes, the four of them are never operating all at once. One is always running near maximum power while the other three are carried as dead weight and reduce flight performance. Thus, at the gain of temperature independence we've compromised flight ability. A design solution for a *strong* flyer would be, of course, to control the temperature of all the engines (or cylinders) and to have all of them operating simultaneously at maximum power. That is the solution that aircraft engineers have arrived at, and it is the solution we find in powerful flyers such as insects. Nevertheless, compromises are necessary at the temperature extremes at which thermoregulation fails. In the neurogenic (but not the myogenic) insect flight motor, one of these compromises involves modification of the sarcoplasmic reticulum.

One of the major factors limiting the rate of muscle contraction in a neurogenic insect flight motor is the rate of calcium uptake following calcium influx to the muscle fibers during contraction. The sarcoplasmic reticuli (SR) are the membrane structures within cells that take up the calcium, readying the muscle for the next contraction. At least in neurogenic muscles, the more SR, the more rapid is calcium uptake, and the sooner the next contraction is possible. One of the adaptations that allows neurogenic muscles to have very rapid contractions is an extensive SR system, which provides many sites for calcium uptake. SR elaboration has been found in the neurogenic wing muscles in some male katydids. These muscles contract at the extraordinary rate (for neurogenic muscles) of over 200 times per second while the insect rubs the wings together (at that frequency but at very low amplitude) during singing. A muscle cell packed full of SR would not produce much power, however, because the SR takes up space needed for contractile elements.

Most insects cannot afford the option of extensive amounts of SR, because generally the power output per wing beat would be too small. One apparent exception is the male *Operophtera bruceata,* a tiny winter-active moth that flies in New England in late fall and early winter. (The females are wingless and the males fly very weakly, in part by having very broad but light wings with which they can drift in a breeze.) *Operophtera* are extraordinary in that they fly even with a muscle temperature near 0°C. We do not yet know how their muscle tissues are biochemically and structurally organized to contract at such low temperatures. But part of their solution is probably the same

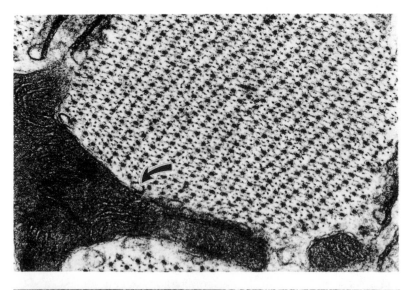

Muscle cells may vary in structure depending on the insect's thoracic temperature during flight. In the sphinx moth, *Manduca sexta (top)*, which flies with a thoracic temperature at least 37°C, the muscle cells contain only a little sarcoplasmic reticulum *(arrow)*. The muscle cells of *Operophtera bruceata*, whose thoracic temperature may range from 2 to 20°C during flight, have large bundles of sarcoplasmic reticulum. (Courtesy of J. H. Marden and R. L. Anderson.)

adaptation found in other very fast but weak muscle—namely, an extensive SR—suggesting that at least one of the limitations to contraction at low temperature is calcium uptake.

Given the many different kinds of insects and flight motors, we might also suspect that power output varies greatly in different species. Because of muscle's high temperature sensitivity, however, it is not obvious to what extent possible differences in power output are due to differences of body size, temperature, or the power capacity of the flight

Males of the winter-flying geometrid moth, *Operophtera bruceata,* may fly (weakly) with a muscle temperature near 0°C. Females *(lower right)* are wingless, and their thorax is filled with eggs rather than flight muscle.

motors *per se*. For example, a bumblebee's flight motor may be more powerful than that of an *Operophtera* moth because its "engine" is large relative to the bee's total mass, because it runs at near 40°C rather than perhaps at 10°C, or because it has design features that make it more powerful than another insect's, weight for weight, at any temperature.

Robert D. Stevenson from Boston University and Robert K. Josephson from the University of California at Irvine have worked to untangle this complex puzzle by separating or isolating the insect flight motor from the rest of the "vehicle" that it powers. These physiologists have put the isolated motor "on blocks" in the laboratory by anchoring it in wax. Such preparations were hooked up to a transducer, and both force and frequency of muscle contractions were then measured. In the insect, the electrical control system is located in the thoracic ganglia, and in the isolated laboratory preparation the signal for contraction can be given by an electric stimulator. Stimulus frequency and hence contraction frequency then can be controlled while the temperature of the flight motor is determined by the temperature surrounding the preparation.

Using these methods, Stevenson and Josephson compared the performance of the stripped-down flight motors with data of the "real-life" performance at different temperatures in the hummingbird-sized sphinx moth, *Manduca sexta* (the larva may be more familiar to most of us as the tomato or tobacco hornworm). They showed that the temperatures at which moths are unwilling to initiate the neural commands to shiver for warm-up before flight correlate closely with the low temperatures at which their muscles don't contract at all. At slightly higher temperatures neural commands were, under the right motivation, given to start shivering, at which point the muscles started to contract slowly, and then at increasingly higher temperatures they contracted more rapidly. Work or power output is a product of both power per contraction times contraction frequency. Slow contractions at low temperatures may be quite powerful, but work output is increased by increasing contraction frequency, up to a point. In the strapped-down moths, of course, the experimenters could override commands from the nervous system to the muscles and supply their own commands in order to observe the effects on muscle-power output at various temperatures and rates of contraction. Interestingly, once contraction frequency reached a high enough level (by imposed electrical commands), power output again declined as the very rapid contractions each produce less and less power. There is an optimum contraction frequency at which power output peaks, and this optimum varies enormously with temperature.

In *Manduca sexta,* the power output during shivering extrapolates to zero at or near 10°C, and warm-up occurs spontaneously generally only at night, at temperatures above 10°C. But once shivering warm-up is started, it proceeds until muscle temperature reaches about 35°C, when the power output of the muscles finally reaches that achieved during free flight. Thus, the behavior of the moth corresponds nicely with the performance predicted from the isolated flight motor; data from the stripped-down flight motor could be used to predict when moths can or should begin to warm up and why they might shiver until they heat up to at least 35°C.

There are, however, numerous other moths—such as most microlepidoptera and geometridae (inchworm moths) and some ctenuchids and arctiids—that are small or weak flyers and that don't heat up but fly at low air temperature. They fly at muscle temperatures much lower than even those at which the large-bodied, small-winged (and hot-blooded) sphinx moths generate zero power. Evolution has acted strongly to tailor the flight motor's capacity for maximum power output for much lower ranges of operating temperatures. For example, the aforementioned moth *Operophtera bruceata* can gain sufficient power to fly at 0°C. Nevertheless, its capacity to do so is only partially due to muscle physiology, and one must search for other design features of the whole vehicle.

The basis for the evolution of differences between species arising from a common ancestor is variation among individuals. Variation was present in the past, and for many traits variation is maintained even now. For example, in his studies of sulphur or *Colias* butterflies, Ward

B. Watt from Stanford University determined that the gene locus for phosphoglucose isomerase, one of the enzymes involved in energy metabolism in these butterflies, changes in allele frequency with season and habitat temperature. This suggests that natural selection is occurring even over very short (that is, seasonal) time spans. The different enzyme alleles have different thermal stabilities, and heterozygotes are thought to have an advantage in an environment of rapidly fluctuating temperatures inasmuch as the individuals heterozygotic for this locus fly over a range of temperatures broader than the range individuals of other genotypes are able to fly over.

When a seasonally changing temperature environment, which is the rule, can select for heterozygosity, then one might expect that an environment of constantly high or low temperature, which is the exception, should lead to the fixation of appropriate genotypes. Hence, selection in terms of gene-frequency changes would not normally be present for our inspection in more constant environments, where appropriate genotypes would already have been selected long ago to adapt to the average temperature. Nevertheless, different life stages of insects may have different biochemical profiles reflecting different thermal and energy demands. For example, in honeybees *(Apis mellifera)* the activities of enzymes involved in flight metabolism increase manyfold in the first four days, when workers normally switch from sedentary to flight-intensive foraging duties. Similarly, James H. Marden, working at the University of Vermont, determined that the optimal temperatures for flight-motor operation in dragonflies *(Libellula pulchella)* change from low to high temperatures, and the enzyme profiles change with them, as the animals change from sit-and-wait strategists when they are young adults to continuous flyers. These examples indicate that insects can or could have considerable biochemical flexibility, and therefore they have the ability to evolve a flight motor that can operate at various temperatures. This means that the *specific* thoracic temperature that is maintained by regulation is "chosen" by evolution probably because it is the temperature most readily regulated for maximum activity over a range of prevailing environmental conditions.

4

Warm-Up by Shivering

During his classic honeybee communication studies, Karl von Frisch noted that bees often interrupted flight for a few minutes when they were returning to the hive heavily laden with nectar. He presumed they stopped "to rest," but we now know they were stopping to work: to raise their thoracic temperature. They most likely stopped flight because it was a cool day and they had cooled convectively. Bees are able to raise their thoracic temperature by shivering, which works their flight muscles harder than flight itself does. Von Frisch could not have known any of this, because shivering and thermoregulation by individual insects was unknown in the 1960s, nor would shivering have been externally visible in bees even if one had looked very closely; even if heat production were measured, shivering would still likely have gone undetected.

All animals produce heat as a by-product of cellular metabolism. As long as an organism is alive, it is constantly producing heat. In *resting* insects, however, this heat production is of such low magnitude that, given the high conductance of an insect's body, it does not affect biologically relevant body temperature. Vertebrate animals that need to elevate their body temperature may, to a limited extent, increase the overall resting metabolic rate of most of the body's cells by increasing the body's output of the hormone thyroxine. Again, insects are far too small to affect body temperature in this way, and in any event they probably have no hormones that increase metabolic rate *specifically* for thermoregulation.

In all endothermic animals there are tissues that are specialized and targeted for heat production. Generally the muscles have a double function, serving for locomotion and heat production. This is a sensible arrangement, because muscles already produce heat as a by-product of doing work, and they are supplied with nerves, which means that heat production can be controlled by the nervous system. Many thermoregulating animals, including humans, increase body temperature with precise, neurally regulated muscle contractions called "shivering." In some animals specialized tissue called "brown fat" is loaded with mitochondria, and it contains none of the contractile elements found in muscle. As a consequence, when this tissue is neurally activated it produces no work but only heat. This process is called non-shivering thermogenesis (NST). NST is most common in neonate mammals who lack muscular coordination for shivering. It is absent in birds who are well-supplied with powerful flight muscles and whose young are incubated by adults.

The idea that we humans might be able to lose weight by somehow inducing NST has long been an attractive and perhaps irresistible lure to the "weight-loss industry." As I write a friend has just brought to my attention the latest commercial effort in this regard. The brochure shows pictures of an obese, heavily clad woman ("Before") and next to her a thin, scantily clad grinning one ("After"). The headline reads "This little pill helped burn 106 lbs. from Lori's body in 2½ months." More photographic documentation is added for emphasis. How did all these people achieve such remarkable results? "Easy, with the amazing formula in ThermoSlim." The ingredients of this pill "stimulate the rapid-fire burning of millions of fat cells" and "aid a person in super high speed weight loss by creating a condition called thermogenesis . . . igniting your internal body heat at a tremendous rate." If we ignore the obvious errors in this copy—fat cells are not "burned" in the live body and heat can't be ignited—the pill's supposed effect on the body's physiology is a mystery, and it is apparently a trade secret. It is not surprising that a sales spiel masquerading as science can be disseminated in brochures, but there have been numerous claims in reputable scientific journals that NST acts as a mechanism of endothermy in insects, even though in no case has the burning of body fuels or an elevation in body temperature ever been demonstrated in the *proven absence* of muscular contraction. NST was only inferred.

In insects all the internal heat that affects body temperature is produced in the thorax, which is packed nearly solidly with the flight muscles. Fat cells are located only in the abdomen, and insects have no "brown fat" or other tissue specialized for heat production, as do some mammals and even some predatory fish that have degenerated and retooled some eye muscles specifically for heat production to enhance vision. It makes little evolutionary sense for an insect to sacrifice any of the power of its flight muscles for heat production when those muscles can already produce heat—and do so superbly—by contracting. With very common current laboratory methods one can measure impressive heat production in some insects, such as bees, but measuring the concomitant muscle contractions in such a tiny animal is often difficult. Shivering warm-up involves a number of sophisticated and intricate mechanisms, some of which have only recently been elucidated. As I will show later, some bees possess, relative to other insects, superior sophistication of shivering and exquisite damping of mechanical vibrations. Shivering by honeybees and bumblebees resembles more the idling of a finely tuned Ferrari than the sputtering of an old clunker, if we consider some loudly buzzing flies and moths whose wing vibrations are visible the "clunkers" of the insect world.

But why shiver at all, if heat is produced already as a by-product of the flight metabolism? To answer this question at the simplest level first (we will expand at length in later chapters), let us consider a sphinx moth that is endothermic in flight. Because of its large muscle mass, a tobacco hornworm sphinx moth (a species studied perhaps more than any moth) unavoidably heats up in flight. It cannot dissipate its metabolic heat to keep its body temperature the same as ambient temperature. Insects of similar or larger size than the sphinx moth are therefore *forced* to operate their flight motor at elevated temperatures. One can reasonably suppose efficiency should be maximized at the muscle temperature that is experienced in *flight,* when maximum power output is required. Large, fast-flying insects must therefore warm up *before* flight to achieve the body temperature they will need once they are in flight. The whine of the syrphid fly (some species of flies are noisy shiverers, unlike bumblebees) you hear before it shoots up from its shaded branch into a shaft of sunlight is the sound of its flight engine revving up as contractions of the powerful flight muscles gradually pick up speed and power. At low enough temperatures the muscles can't contract at all, but once shivering proceeds and the muscles get warmer, they can produce faster and ever more powerful twitches until optimum flight temperature is achieved just prior to takeoff.

Mechanisms

Like the maintenance of an elevated body temperature by internal heat production in flight, physiological warm-up is found in all large, active flyers among the dragonflies (Odonata), moths and butterflies (Lepidoptera), katydids (Orthoptera), cicadas (Homoptera), flies (Diptera), beetles (Coleoptera), and wasps and bees (Hymenoptera). That is, it is found from some of the earliest forms, the Odonata, to the most evolutionarily highly derived, the Diptera, Coleoptera, and Hymenoptera. It is *not* found in the small and therefore non-endothermic members of the same groups. Since no insects shiver except those that then also heat up from flight metabolism, it seems reasonable to conclude that the evolution of shivering behaviors is related to the evolution of flight but is *un*related to the insects' place on the phylogenetic tree.

It might be posited that a or the common ancestor of insects originally shivered and that shivering thereafter evolved once so that those descendants that didn't need it lost the capacity to shiver.

The "flyer" dragonfly, *Anax junius,* shivering during pre-flight warm-up.

Alternatively, it is possible that shivering capacity is not a direct inheritance from a common ancestor and that instead it evolved independently hundreds or thousands of times in all of the diverse species or groups of closely related species. The data fit the second hypothesis. First, as far as we know the present-day (weak-flying and small) mayflies don't shiver, even though they bear the closest resemblance to the line leading to all insects. Second, as determined by the late Ann E. Kammer of Kansas State University, who perhaps more than anyone else elucidated the neurophysiological mechanisms of shivering in insects, there are subtle physiological differences in the shivering response even in very closely related insects.

While working with five species of sphinx moths collected in California (*Celerio lineata, Manduca sexta, Sphinx vashti, S. perelegans,* and

Smerinthus cerisyi), Kammer recorded the neuronal activation patterns of flight muscles by means of electrical recordings from wires placed into the muscles. In all five species wingstrokes during pre-flight warm-up were produced by synchronous contractions of groups of muscles that normally contract alternately in flight. That is, the main wing-depressor muscles, the dorsal longitudinal (DLM) muscles, were excited simultaneously—in other words, in synchrony—with the dorso-ventral (DVM) wing-elevator muscles. However, three direct muscles (which pull the wing directly in flight not for power but for flight maneuvering)—the subalar, basalar, and third axillary muscles—were usually excited out of phase with the DLM and in phase with the DVM. However, details of motor patterns varied from one species to the next. In *Mimas tilia,* for example, the subalar is fired synchronously with the DLM rather than with the DVM. The differences suggest that a variety of warm-up patterns evolved even within these closely related moths as slight modifications of a common flight-motor pattern.

The neural activation pattern of the thoracic flight muscles needs to be and is already very labile for flight control, and to add shivering

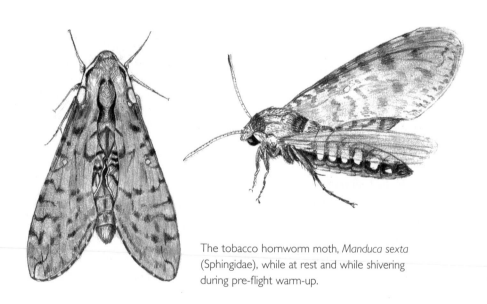

The tobacco hornworm moth, *Manduca sexta* (Sphingidae), while at rest and while shivering during pre-flight warm-up.

when you already have flight behavior is probably a very minor evolutionary step. Physiological warm-up in its most basic form is like the idling of an engine; the engine "evolved" to propel the car, not to warm it up. Once present, the heat-producing flight muscle system required only a slight modification of neuronal activation patterns and, in the more sophisticated models, also the addition of the biological equivalent of a clutch—a mechanism to disengage the wings in the same way that an automotive clutch disengages the car's wheels.

Various levels of sophistication of the shivering response are found, however. Some insects, such as dragonflies and moths, demonstrate a simple form of shivering. They have visible external wing vibrations that were originally called "wing whirring" and were once thought to pump air into the animal before the true function was elucidated. The slight vibration of the wings or "wing whirring" one sees during shivering by butterflies, moths, and dragonflies results simply from slight inequalities in the timing or the strength of contraction of the various wing muscles.

Since each nerve impulse, or group of closely spaced nerve impulses, causes a muscle to contract in a twitch, the main difference between shivering and flight behavior is that in one the wing muscle contractions are nearly synchronous whereas in the other they are alternate. During shivering warm-up it is as if all cylinders of a car engine were caused to be ignited in synchronous bursts.

Ever since the wings of insects evolved from mere paddles used to stir the water to increase aeration, their mechanical coupling to the body has become more complex. The highly derived insects, including Diptera and Hymenoptera, have a more complex wing hinge than Odonata and Lepidoptera, and in most of these insects this hinge can uncouple the wings from the muscles driving them to such an extent that one can neither see the movement nor hear the vibration of the wings; that is, they superficially mimic non-shivering thermogenesis. Researchers can detect mechanical vibrations only with sensitive instruments applied at the right place.

As already suggested, the zenith of the shivering response of any hot-blooded animal (vertebrate as well as invertebrate) belongs to some bees. Honeybees and bumblebees have a physiological sophistication either not existing or not yet observed in other insects, and they exploit shivering behavior to an unprecedented extent and in a variety of ways. Like flies and beetles, bees are "myogenic" flyers in which the wing-beat cycle runs in part on automatic; as the downstroke muscles contract they stretch the upstroke muscles, and this stretching *by itself* causes the upstroke muscles to contract. The downstroke of the wing therefore automatically causes the upstroke muscles to contract and vice versa, in a repeating cycle that is sparked by neural commands that are now sent at a much lower frequency than the wing beats and are no longer specific to a single wing beat. (This system permits some of the smallest insects, such as midges, to achieve the unprecedented coordination required for wing-stroke cycles of over 1,000 beats per second.) But the stretching of opposing muscle groups that maintains the myogenic contraction cycle can only occur if the wings are actually beating—namely, during flight. When the wings are not in use, as when they are folded back dorsally and the clutch-like wing-hinge is engaged, then the stretching of one muscle group is insufficient to trigger their contraction. The ability of myogenic insects to shiver had long been a mystery, if not dismissed outright as an impossibility.

The solution has only recently come to light, mostly through the elegant work of Harald Esch, Franz Goller, and Ann E. Kammer. When muscles are *not* stretched to their full extent by their antagonists, as they are during the full wing-beat cycle, then they must be neurally stimulated for *each* contraction. After neural stimulation, they contract by ordinary twitches. In myogenic flyers like bees, the neural activation patterns of the flight muscles are thus markedly different between shivering and flight. During flight, the up- and downstroke muscles are activated only occasionally and almost randomly with widely spaced, single nerve signals (called "action potentials" because an electrical *potential* is generated across the membrane of a nerve cell). During shivering, in contrast, the two sets of muscles are activated *synchronously*. (Synchronous activation also occurs in the neurogenic shivering warm-up of moths and dragonflies.) Also, while in most Lepidoptera only one or two action potentials initiate each twitch of any one thoracic muscle during shivering, in bees the shivering muscles are activated by long *bursts* of synchronous action potentials per muscle contraction. These bursts cause a long continuous contraction and thus dampen vibrations even more than precisely timed twitches of opposing muscles could accomplish.

Very vigorous shivering in bees is physically dampened even more by yet another mechanism. One of the two sets of opposite-acting muscles is activated (and hence contracted) slightly more than the other. Since the opposing muscles act like weights forcing down each side of a seesaw, the added force on *one* set of muscles prevents the "seesaw" from working (and the wings from "vibrating" back and forth). As a result of this "silent" shivering, the bees have opened up a new communication channel; they can use sound. If they could not silence themselves during shivering, then at low temperature their colonies would resound with a cacophony of noise that would make it difficult to relay acoustic messages.

Other Functions of Shivering

Shivering, or very close physiological variations of it, can result in a buzzing sound. African tsetse flies, *Glossina morsitans* (the famous carriers of sleeping sickness), are especially noisy during warm-up. It has long been reported that they whine before takeoff and that the whine becomes increasingly higher-pitched throughout warm-up, as though it were a signal in communication. We now know that the higher pitch results from the increasing temperature of the wing muscles: as the muscles become hotter they contract faster (and so produce a higher-frequency, higher-pitched sound); by contracting faster they produce even more heat and contract faster still, etc.

To my knowledge, buzzing occurs only in Hymenoptera and Diptera, and it involves partial wing opening from their normally locked-in, folded position and partial relaxation of the thoracic muscles. (In shivering *per se* the wing movements are usually completely damped because the opposing muscles are continuously contracted against each other in a near tetanus.) Buzzing in these species indeed functions as a warning signal. For example, when one grasps a bee or a wasp (or a fly that mimics either), the insect buzzes loudly. This is a signal to let go, and it works especially well on those animals that have heard it before when they got stung.

As indicated, a second evolutionary offshoot from shivering in bees is sound for social communication. Insects have myriad ways of producing sound—some cockroaches hiss, many beetles have friction files in the neck, acridid grasshoppers rub leg against wing, crickets and katydids vibrate their wings, cicadas vibrate their "tymbals" (a drum-like membrane), and some *Drosophila* rub the abdomen on the thorax to make courtship songs. The various mechanisms give ample evidence of the many possible origins of insect sound production, but they all suggest that the *buzzing* in flies and bees is derived from shivering, and not vice versa.

A third evolutionary offshoot from shivering behavior involves pollen foraging. In many flowers, such as tomato flowers, pollen is encased in closed or partially enclosed anthers and is available only to bees who harvest it by grabbing the flowers and shaking them. The shaking is accomplished by buzzing. Vibrating by making shivering-like contractions of the thoracic muscles, the bees loosen the pollen and then collect it as they do other pollen. This behavior also keeps the flight motor heated up. It was probably secondarily derived from shivering for warm-up since bees normally shiver anyway (to keep flight-muscle temperature elevated) when they are perched on flowers at low ambient temperatures.

Energetics

Warm-up of an insect flight motor requires the expenditure of considerable energy, and energy is usually at a high premium in nature. An insect therefore faces a crucial decision on whether or not to initiate warm-up. Ambient temperature (or initial body temperature) has a huge effect on warm-up costs. At the extreme of the negative energy balance sheet is an insect that attempts to warm up at an environmental temperature that is too low for it to reach minimum takeoff temperature. Suppose, for example, that air temperature is 10°C and the insect is capable of generating a temperature excess ($T_b - T_a$) of only 25°C but needs a flight-motor temperature of 36°C for takeoff. When such an insect *begins* to shiver at an ambient temperature of 10°C, almost all of the heat it generates acts to increase body temperature. Since biophysics dictates that its rate of passive heat loss (by convection to the air) is proportional to temperature excess, however, the shivering insect loses more and more heat as its body temperature begins to rise. The animal *continues* to get hotter only because as it gets warmer it also is able to produce more heat (because its metabolic rate increases). A continual increase in heat production can make up for and exceed the increase in convective heat loss. Eventually, however, body temperature must stabilize because passive convective cooling becomes too great and heat production reaches a limit. If our hypothetical insect (which can generate a maximum temperature excess of 25°C) started to shiver at 11°C, it could reach its minimum takeoff temperature of 36°C, but if it had started only 1°C lower, at 10°C, it could shiver indefinitely (or until it exhausts all of its fuel) only to stabilize its temperature at 35°C and never quite reach 36°C. In other words, an insect trying to warm up at an ambient temperature that is slightly too low is like an airplane that taxies down the runway with almost, but not quite, enough power to lift off. A little bit of difference can make *all* the difference.

We do not know if insects make calculations on whether or not to warm up, but they undoubtedly have rules of thumb that help them maintain energy balance. For example, most insects readily start to shiver at temperatures that are moderate for the environment, but they become increasingly hesitant to shiver as they reach the lower end of the range. The use of internal temperature set-points to "trigger" warm-up does not fully explain their behavior, because in some species (such as bumblebees) size range may vary about seven-fold (and rate of heat loss varies several-fold) between genetically nearly identical individuals. It is highly unlikely that genetically determined set-points would vary by individual body size (which in the bees is very labile). The smallest individuals can fly only at high air temperatures because they cannot warm up at ambient temperatures that are quite suitable for larger individuals.

The "start" signal for shivering is given to the flight-motor muscles by electrical impulses analogous to the electrical current that fires the spark plugs and ignites fuel in the cylinders in a car. After shivering begins, the next problem is to decide at what rate to proceed. Again, there is an optimal energetic solution, namely, to warm up *as fast as possible*. The whole object of warm-up is to store the heat that is generated. This is achieved by heating the flight motor as quickly as possible, so that the heat loss by convection is minimized over the whole warm-up cycle. We know that the "all-out" warm-up rule is followed by insects because warm-up is almost always much faster at high than at low temperatures, inasmuch as hot muscles can accomplish more work than cool ones. In summary, the insects' decision to warm up is made with caution, but once made activity proceeds at full throttle and without a break. I've generally tried to follow their example, because as a rule of thumb, the efficacy of their strategy applies to the efficient and economical performance of almost any task.

Aside from maximizing shivering rate during warm-up, there are several other means that insects use to advantage that minimize the wasteful loss of energy by convection. First, they reduce the energy costs of warm-up by heating only the flight motor, and not the rest of the body, although this is less a matter of choice than a consequence of their anatomy. Endothermic moths and dragonflies in particular show almost no elevation of abdominal temperature throughout shivering warm-up. That is, they concentrate all of the heat in the flight motor where it is needed most. They do this by stopping or greatly slowing down their heart activity, which reduces blood circulation and the heat transfer associated with it. Second, I suspect that they may seek microenvironments in which to warm up where convective heat loss is reduced.

Social insects have an inherent warm-up advantage because of their colonies. At −50°C in an Alaskan winter, when it is too cold to start cars in the morning, there may nevertheless be traffic on the roads. That is because the vehicles were either left sheltered in heated garages or else left running all night. Similarly, bumblebees and wasps are routinely out in the field on mornings when the temperature is near 1–2°C, too cold for their muscles to even begin shivering. They are able to start foraging at these temperatures because their muscles were kept warm enough to start shivering in the shelter of the warm nest; once they have started foraging, they usually keep their flight motor continually running even when they stop at flowers.

I conclude that while all endothermic birds and mammals shiver, insects have perhaps by far the most sophisticated mechanisms of keeping warm. They undoubtedly began shivering far earlier in the evolutionary history of life than any vertebrate animal. Indeed, they have even evolved to *refrain* from shivering when it is advantageous to maintain a low body temperature, which for many of them is often if not most of the time.

5

Warm-Up by Basking

An animal that can shiver can begin activity by warming up at almost any time. But it pays the price of having to invest energy to fuel the process. On the other hand, the energy of the sun is free, and an insect could potentially perch in sunshine and heat up "passively." The problem, however, is that there is never any assurance that sunlight will be available. It is also difficult to avoid predators while perching in the open. Thus, both methods of warm-up—shivering and basking—have costs and benefits that vary under different ecological circumstances. The balance between the advantages and disadvantages depends greatly on body size.

The temperature of the flight motor of a 100 mg syrphid fly that parks in a sunfleck may increase by 10°C or more per minute, and the insect can soon fly without having expended any energy in shivering. On the other hand, if you try to warm up your car's engine by parking in the sun at 8 A.M. and wait for it to warm up, you might have to wait until noon to leave for an important appointment. On a cloudy day you might not be able to leave at all. The fuel you save by "basking" rather than idling the motor for a minute or so is not worth either the time you'd spend waiting or, given the unreliability of the weather and the shortness of the day, the risk of never heating up the engine. But in many insects the large relative surface area because of small body size that is a disadvantage for retaining endothermic heat is simultaneously an *advantage* for gaining heat in sunshine. It allows them to heat up quickly because every point in the body is close to the heat source.

Like warm-up by shivering, warm-up by basking occurs in all major orders of insects that have fast flyers large enough to heat up from their flight metabolism. In its simplest form, behavioral warm-up is merely heat-seeking. When an insect happens to crawl out of cool shade into sunshine, it may feel comfortable and stay there. Additionally, slight shifts of body posture can enhance solar heating. Such responses can, however, only marginally be called "basking." In most insects that bask, basking involves more highly coordinated and specialized behavior that is often absent in closely related species.

A basking insect usually takes specific postures that simultaneously maximize solar input and minimize convective heat loss. Heat input is maximized by exposing the maximum surface area to the sun, while convective heat loss is minimized by using body parts (such as the spread wings) as baffles to retard air movement around the body.

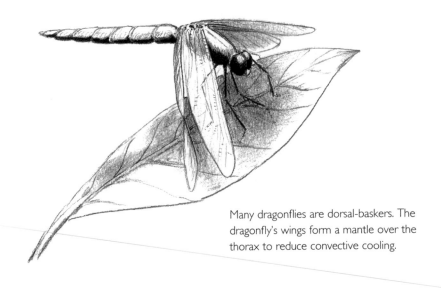

Many dragonflies are dorsal-baskers. The dragonfly's wings form a mantle over the thorax to reduce convective cooling.

Orienting the body parallel to the air stream (as a wingless insect might do) would reduce the effect of convective cooling, but orienting the body perpendicular to the sun's rays to facilitate heating should take precedence, because no heat loss can be minimized until heat is first gained. Grasshoppers, beetles, and flies use these basking methods, and for some dragonflies and butterflies the wings are especially important in warming up.

Basking Postures

Basking was well-known in insects long before it was studied in vertebrates, principally lizards. The first observations were made on butterflies, over a century ago, and perhaps because of their colorful and conspicuous wings, basking is still most well known in them.

Four behaviorally distinct types of basking have been described in butterflies, although some (tropical) butterflies don't bask at all. In one type, called "lateral basking," the butterfly closes its wings dorsally and then tilts to present either the right or the left wing and body surface to the sun. The lower portions of the wings wrap around the body and even touch it, and warming these lower wing portions in sunshine causes heat to be conducted directly through them into the body. On almost any mountain meadow you can generally see butterflies stopping in this tilted position after flying a few seconds, because at some temperatures small butterflies will, as the honeybees described in the last chapter, cool convectively as they fly. They cool sufficiently in flight that they need to stop to heat up. This type of intermittent basking associated with flight is practiced by most species of butterflies at the low temperature range in which they are active. Some of the small-bodied species, such as satyrids and lycaenids, that cool much more quickly than larger insects also engage in this behavior, even on relatively warm summer days.

Many species of very small-bodied butterflies, primarily pierids and some lycaenids, that commonly fly in breezy mountain meadows bask by opening their wings partially in a *V* so that the body is in direct exposure at right angles to the sun. The sun's rays intersect the body surface at the bottom of the *V*. Although the heating of the body can be accounted for by sunlight striking the exposed body directly, the wings have an enormous role in basking. In bodies as small as these butterflies are, the slightest breath of moving air causes considerable convective cooling. Raising the wings into the *V* accelerates heating rates because convective cooling from ambient air movements, which would normally cause body temperature to plummet, is greatly retarded by the wings; in other words, the wings serve as convection baffles over the dorsal body surface. (It has also been suggested that the flat wing surfaces focus heat onto the body, as by a lens, and also that the wings work like commercial solar panels shunting heat by circulating hemolymph. Undoubtedly both processes exist in the realm of physics, but available data do not show biologically relevant temperature changes. They are more in the order of using a flashlight to heat your house when you already have a furnace).

Still a third type of basking is observed in many dozens of species of the larger nymphalid butterflies, such as the well-known mourning cloak, *Nymphalis antiopa,* of both Europe and North America. This butterfly and a number of its close relatives overwinter as adults and fly even before all the snow melts in the spring. Overwintering nymphalids commonly bask on open areas on the ground or on tree trunks and rocks. To bask they face away from the sun and orient the dorsal body surface at right angles to the sun's rays, as do the V-baskers. Here too the primary heating is caused by direct sunshine striking the body surface, and the wings again play a vital role in basking. In this case, however, the wings are held *down* and to the sides, thus reducing air movement on the ventral area of the butterfly. While this posture reduces convective cooling from the undersides, it also traps warm air rising from the sun-heated substrate. Dead butterflies fixed in this basking posture heat as rapidly as live ones do, and cutting off the wings of model butterflies and fixing them in place, or substituting artificial wings (made out of paper), has little effect on body temperatures. The wings thus do not act analogously to our

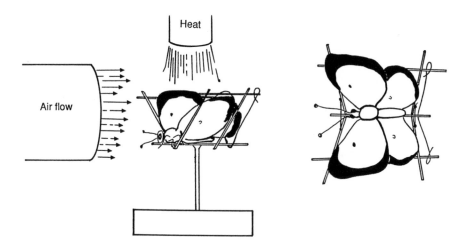

Heat

Air flow

The experimental set-up used to study the role of wings in convective cooling and possible reflective heating. A dead butterfly's wings were raised or lowered while simultaneously subjected to heat from a light from above and to convective heat flow by air flow from the left. The butterfly could be turned to face the air stream or to be lateral to it. Changing the wing angle to vary the amount of light intercepted or its direction had no effect on the butterfly thoracic temperature, provided the butterfly was facing the air stream. But changing the wind direction had a large effect on thoracic temperature, if the wings were raised.

commercially available solar panels, in which heat is gained in one area and transported to others by circulating fluid. Fluids do circulate in insect wings, but the volume of fluid circulating in butterfly wings has been shown to be insufficient to have any practical effect on body temperature; stained hemolymph becomes distributed in the wings in a matter of hours, not seconds.

Dorsal-basking butterflies do not, however, generally hold their wings still, either up or down, at the same angle. Instead, soon after beginning to bask, most dorsal baskers, especially nymphalids, slowly and rhythmically move their wings up and down in "wing pumping" movements. Helmut Schmitz and Lutz Wasserthal at the University of Erlangen have found a possible explanation for this

The orange sulphur butterfly, *Colias eurytheme (left)*, lateral-basking; the closed wings are tipped to the side to present a broad, lateral surface to the sun.

The eastern blue butterfly, *Everes comyntas (right)*, is a dorsal-basker. The raised wings reduce convective heat loss over the dorsal body surface; they have been hypothesized to aid in "reflective" heating.

A skipper butterfly (Hesperidae), in another dorsal-basking posture *(left)*. Its forewings are raised to reduce convective heat loss across the dorsal body surface, and its hindwings are lowered to reduce ventral convective heat loss.

The comma butterfly, *Polygonia comma (right)*, hugging the ground in a dorsal-basking posture. The wings prevent or reduce convective heat loss from the ventral side.

until recently puzzling behavior. Their studies suggest that the distribution of heat in the body is involved; various body parts are subject to very much different heating and cooling rates and very different equilibrium temperatures in sunshine are achieved. For example, the thorax of a *Pachliopta aristolochine* (Papilionidae) requires approximately 8 minutes to heat to an equilibrium temperature (about 42°C) from 25°C in sunshine, while the spread wings heat to above 46°C in just 30 seconds. The antennae (because of their very large surface area relative to their mass and exposure to the surrounding air) barely heat up at all in sunshine, hence thermal sensors on them measure air temperature. Furthermore, experimental stimulation of thermoreceptors in the wing veins (with a light spot to focus heat) evokes wing-closing movements that then quickly cool the wings, suggesting that the wing movements may function to protect the basking butterfly from the damage of overheating its wings when they are continually exposed to the sun. Since the thermosensors in the wings seem to react to the steepness of the heating rate, they could warn the butterfly when wing-"frying" is imminent. Although the butterfly's eyes are in part responsible for sensing its orientation to sunshine, it is also possible that the wing thermoreceptors might make it possible for the animal to irradiate both wings equally, and thereby to achieve and maintain the dorsal basking posture.

The fourth basking behavior is confined to a group of small-winged and thick-bodied butterflies, the skippers (Hesperidae). Skippers bask dorsally and simultaneously raise their front wings and lower their hind wings. By extrapolation from other butterflies, it seems likely that the raising of the front wings retards heat loss by convection across the top of the thorax while the lowering of the second set of wings reduces convection across the bottom. Skippers, unlike lateral baskers, for example, also commonly warm up by shivering, as does the mourning cloak and most other dorsal baskers. In general, however, different species of butterflies either bask or shiver. Only some do both.

Basking Costs

Bees do not bask. They keep warm by shivering instead. If you watch their behavior at flowers, however, it almost immediately becomes apparent why they don't bask. Basking requires the insect to orient itself in a specific position to the sunshine, and for that it must remain still. Bees can't remain still and at the same time maintain high foraging rates, any more than the mail carrier can take the time to park her car perpendicular to the sun's rays at each mailbox so as to take advantage of solar heating.

To both bees and mail carriers, time is of the essence, and proper orientation to the flowers and/or mailboxes is of far greater relevance than an advantageous orientation to the sun would be. Like the mail carriers undeterred by the weather "from the swift completion of their appointed rounds," bees just keep on humming, they don't stop for sunning. Most of the flowers from which they forage require specialized behavior—bees open lock-like obstacles to reveal hidden nectar,

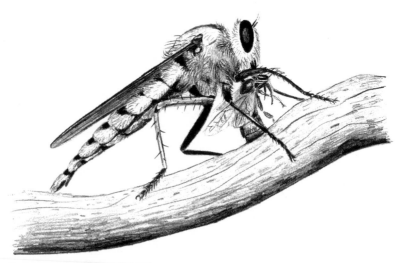

The robber fly, *Promachus giganteus* (feeding on a honeybee). This insect stays warm by basking while perching so as to be able to fly fast to snatch flying prey.

they stand upside-down, they make "buzz-runs" to collect pollen over platforms of anthers—that would infere with maintaining a specific body orientation to the sun at the same time. Many flowers themselves orient to the sun, so the bees cannot avoid some exposure to the sun, which is probably useful to them. Many flowers are shaded, however, even on sunny days.

Basking is ideally suited for sit-and-wait strategists, whether an insect is searching for a mate or a meal. Syrphid fly males and some male butterflies that wait in a sunfleck on a cool day and that are instantly ready to chase a stray passing female are prime examples of insects expected to bask. Among the sit-and-wait predators are also robber flies (Asilidae) and percher dragonflies that wait for prey to come near them.

One may safely conclude that relative to *shivering,* basking or selecting a warm micro-habitat yields an enormous energy saving for an insect trying to warm up. On the other hand, the benefit may be won at a great cost in terms of elevated resting metabolism (see Chapter 6) and predation. In sum, basking is a savings plan that many insects need but simply can't afford.

6

Cooling Off

Other things being equal, large animal bodies cool more slowly than small ones. That's because the ratio of surface area to body mass or weight is lower for large bodies than for smaller ones, and having relatively less surface area from which heat may be dissipated (lost from the body to the environment) lowers the rate of heat loss. For two reasons, then, a relatively small amount of heat production per unit mass may lead to a rapid buildup of heat in a large body: the larger the muscle mass, the more heat is produced; and the smaller the relative skin surface, the less heat is lost.

This has practical application in the design of engines, both mechanical and biological. A small car can easily scamper through the blistering hot desert near Tucson with no special cooling mechanism, whereas an 18-wheeler truck would quickly overheat if it were not equipped with a cooling system. Neither vehicle is designed to run at an optimum temperature; that is, the truck is not made to run more efficiently at high air temperatures than at low to avoid overheating, and a small car is not designed to run less efficiently (by consuming more fuel) at low air temperatures in order to remain hot. To a large extent, the same applies to insect flight motors; the energy expenditure of flight muscles does not decrease during locomotion at higher temperatures so that, by consuming less fuel, they will not heat up so much. Instead, overheating is prevented by an increase in heat loss at high temperatures, and that is done by incorporating cooling mechanisms.

Heat loss by evaporation of water is the most powerful cooling mechanism for living organisms, and it is widely used by animals of large body mass, such as cattle, humans, horses, and species of large antelope. But no engineer has so far designed cars or planes to carry huge water supplies specifically for thermoregulation at high temperatures. Nor are insects designed that way. Instead, most of the heat loss in both mechanical engines and insect flight motors ultimately results from convection.

Harnessing Convection

The rate of convective heat loss from a body is determined by the conductance of the body, or its intrinsic rate of heat loss (that is a function of body size, shape, and insulation). Conductance, in turn, is a function of the wind speed (a body cools off more quickly in wind than in still air—meteorologists call this the "wind-chill factor"). However, no convective heat loss is possible, regardless of conductance, if body and ambient temperature are equal, and at any one conductance and wind speed the amount of heat loss is directly proportional to the temperature difference between the body and the ambient or surrounding medium, or $(T_b - T_a)$.

Small insects that are endothermic in flight are sufficiently air-cooled that they almost never reach the potentially dangerous high-temperature ceiling of near 45°C that is common to most animal tissues at normal atmospheric pressures. (There are organisms living at thermal vents in the ocean depths that, probably because of the intense pressure of the water environment, can maintain their biochemical integrity at much higher temperatures.) These small insects have thus no need of a specialized cooling system: they lose heat effortlessly. Theoretically, larger insects could cool themselves by increasing flight speed, and thus increasing convective heat loss, but flying faster would increase metabolic heat production, which would cancel out the heat loss.

Few data are available to work out the exact balance between heat loss and heat gain during flight, except for bumblebees. Bumblebees are notorious for having short wings, which were long reputed to be inadequate for flight. These insects indeed do drop like a stone if they are not first allowed to warm themselves up for flight. Recently, however, Charles P. Ellington at Cambridge University and Timothy M. Casey at Rutgers University, an expert on insect thermoregulation, made history by flying bumblebees in place against an air stream of variable velocity in a wind tunnel while simultaneously measuring their energy expenditure. The bees were tricked into the illusion that they were speeding along by appropriate visual cues, such as we might see if we watched the belt on a treadmill as we ran on it. Energy expenditure was determined by measuring their rate of oxygen consumption, which can be converted to equivalent food consumption since the bees are strictly aerobic; like other insects they do not incur any oxygen debt. The bees consumed 50–60 ml O_2/g body weight/hour. Put another way, the two researchers pointed out that while a "standard human male" when jogging burns the energy equivalent of about one Mars bar per hour, a bumblebee in flight would need one every 30 seconds for the same unit of body weight.

It was already known that the bees' energy expenditure in flight doubles if they lift honeycrop loads that equal body weight, and that energy expenditure is independent of air temperature. The main result of the experiment on free-flight metabolism was that the flight costs remain independent of flight speed, at least until fairly impressive flight speeds (relative to air movement) of 4.5 meters per second are attained. Bumblebees can, however, probably fly twice as fast as that, and if they push to the limits of their flight speed they will probably need to increase their energy expenditure.

The ability to fly over the wide range of velocities from zero to 4.5 meters per second *without* altering heat production must have thermoregulatory significance, because velocity changes of several meters per second can have an enormous effect on convective cooling in some insects. The cooling rate increases sharply with large increases in velocity, especially in insects that have only short pile, such as *Centris* bees, which are naked or bald, as are many carpenter bees, *Xylocopa* sp.

Despite their large size, carpenter bees appear to be built to lose heat. Unlike most other bees, they chew galleries or tunnels into solid wood (to provide a safe place for their brood). They have large, flat heads to anchor their thick mandibles and heavy chewing muscles, and their heads fit like a cap onto the front of the thorax. Head and abdomen are smooth and bare, and the thorax is usually also bald. The bees' bare heads and abdomen, with their large surface areas, are excellent radiators for convective heat loss. When a carpenter bee is flying at 12 meters per second, the cooling rate of its head is more than ten times greater than the cooling rate of its thorax. Since the head directly contacts the thoracic flight motor, it draws heat from the thorax by conduction, and the head thus serves as a heat radiator. Active physiological heat transfer to both the abdominal and head radiators also occurs by the circulatory system. Under experimental conditions in a temperature-controlled room, these bees increase flight speed with increasing ambient temperatures. Conversely, in the field they hover more at low air temperatures, as if to conserve heat.

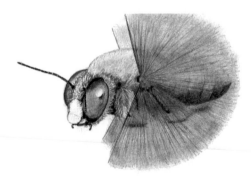

A hovering male anthophorid bee, *Centris pallida*. The bees of this species hover longer at low ambient temperatures than at high, suggesting that they regulate thoracic temperature in part by regulating convective cooling.

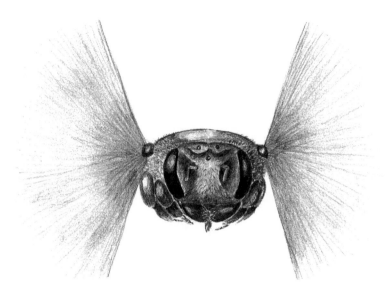

Carpenter bees, *Xylocopa* sp., are large and usually naked bees that live in hot desert. They lose considerable amounts of heat through their broad heads and increase heat loss by flying faster.

Flight is a very effective cooling mechanism, especially if it results in escape from a hot zone. The handsome metallic green or copper tiger beetles (Cicindelidae) use this strategy. These close relatives of carabid ground beetles, with large scimitar mandibles and large eyes, are predators almost the whole world over. They dash across hot sand surfaces as fast as one can walk. But they can also readily take flight. They are active on sunny days on hot sand, where they are sandwiched between heat from solar radiation above and heat from the hot sand below.

To escape this thermal sandwich, tiger beetles harness convection cooling. When they are heat-stressed they first stilt above the hot sand, and if that isn't sufficient to cool them they initiate short flights a few centimeters above the ground, where they encounter air temperatures considerably lower than those near or on the ground. As shown by Kenneth R. Morgan, whose experiments used very small implanted

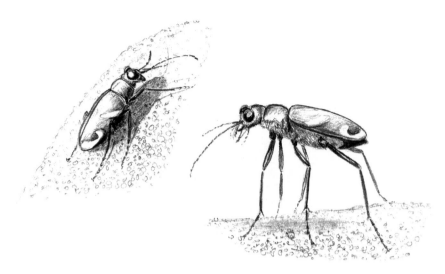

Tiger beetles, *Cicindela hybrida,* are both swift runners and swift flyers who bask *(left)* at sand temperatures near 19°C and stilt *(right)* or fly to cool off when soil temperatures exceed 40°C.

thermocouples on light leads, the beetles initiate a short flight when they start to overheat, and then in a few seconds of flight they achieve a large drop in body temperature. They cool because the gain in heat due to the flight metabolism is less than the increased convective heat loss during flight at a greatly increased temperature gradient above the ground.

Convective cooling by increasing locomotory speed appears to be common in desert insects heated by sunshine, and it is used by some desert ants and by flightless, ground-dwelling tenebrionid beetles, such as the long-legged *Onymacris plana* from the Namib Desert in southern Africa, which can run at almost a meter per second. This speed is fully 20 percent of the running speed of the American record at 100 km—which was set on a solid asphalt surface, not on loose sand. The energy cost of their increased running speed, as indicated by measurements made by George A. Bartholomew and associates of beetles running on small treadmills, is small; running cost is negligible relative to insect flight metabolism in general, but the increase in convective heat loss at this speed is enormous. Fast running could also reduce exposure time to sunshine when the insects are shuttling between thermal refugia, such as patches of shade. Human runners are too large to unload appreciable amounts of their metabolic heat from running by convection, but the size effect on thermal balance is clearly apparent: people of slight build, such as the Nilotic people of eastern Africa, are able to be active under intense heat and work loads, whereas stockier people would quickly succumb from the heat or run out of water. By running, beetles are able to unload, through forced convection, solar heat they receive from above; they thus cool down rather than heat up as a result of their running. Human distance runners are also scantily clad when they compete under a hot sun, but they generally drink fruit juice both for fuel and to replace water lost to evaporative cooling. The beetles, however, have no fluid to spare. For them evaporative cooling is not an option, although when *in extremis* male beetles extrude their aedeagus (penis) for a brief and local cooling respite.

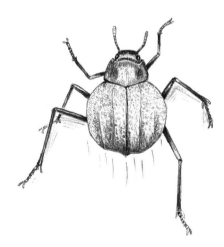

The flightless desert tenebrionids, *Onymacris plana*, are swift runners that may lose heat by convection while speeding along at nearly one meter per second.

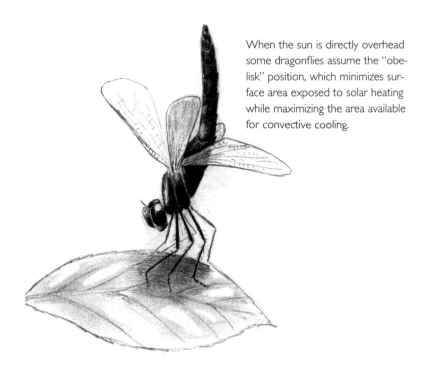

When the sun is directly overhead some dragonflies assume the "obelisk" position, which minimizes surface area exposed to solar heating while maximizing the area available for convective cooling.

The running beetles and running ants have long legs and they may escape some of the desert heat by stilting, but they are then still subjected to the direct solar heat input from above. An earthbound animal might orient its body vertically to reduce heat input from above and below, but few insects can afford to take the vertical body posture because doing so compromises locomotion. Only sit-and-wait predators have it as an available option (unless the insect can also learn to run on its front or hind legs). The vertical position can at least be taken to avoid heating when an insect is perched. And indeed it is found in some dragonflies, where it is called the "obelisk posture."

Peter Wheeler, a physiologist at Liverpool Moores University in England, has recently studied how erect posturing reduces the body surface area that is directly exposed to the intense midday sun and how much water would be saved if the animal had to rely on evaporative cooling. The amounts of water saved by standing erect in sunshine were on the order of a quart or two per day. But Wheeler wasn't studying beetles or dragonflies. He was studying humans—specifically, models similar to Lucy, a fossil representative of an australopithecine that lived three million years ago.

Wheeler's thesis is that our forebears did just as the dragonflies undoubtedly did for millions of years before any hominids were on the horizon. They left the forest to live in the hot African plains, and they held themselves erect on two legs in part to escape the heat of the ground and the direct sun, and thereby to economize on the use of water for sweating. Then they kept on walking, perhaps freeing their hands to make tools. Upright posture also allowed proto-humanoids to become naked—unlike the quadrupeds, which need fur to insulate them from the burning rays of the sun. Nakedness made for more efficient cooling by convection as well as sweating. In the case of the humans, the metabolically most active tissue that is in most danger of overheating is the brain, not the thoracic muscles. When dragonflies take the obelisk position, the thorax is protected by the raised abdomen, which serves as a heat shield. In the humanoids, however, the heat shield of the head and thorax was probably a thick thatch of hair, which is still a valued and often useful possession.

Heat Radiators

A radiator is a device that increases the surface area of a body or object so that more heat can be transmitted to the surrounding environment by convection. In some radiators a fluid with a high heat capacity (like water) circulates by means of a pump and transfers heat from its source to the radiator site for the heat loss. That is how a car engine is cooled.

Insect "radiators" work the same way. Heat may be lost from the surface area of the head or abdomen, and since these body parts already have hemolymph circulating through them, and since nature generally modifies and retools what already exists, these radiators

A hovering white-lined sphinx moth, *Hyles lineata*, sipping nectar. This moth flies primarily at night when days are hot, or it flies in the daytime when nights are too cold. Continuous flight without stopping to cool necessitates the use of the abdomen as a heat radiator for heat loss.

The day-flying "hummingbird" sphinx moth, *Hemaris* sp., sipping nectar. It might be using its front legs to reduce the metabolic cost of hovering, and therefore reduce the rate of heat production.

could potentially be used for controlling thoracic temperature. In most (but not all) flying insects, however, head temperature is generally automatically high even at low air temperature, because heat is passively conducted to the head from the hot thorax. Heat loss through the head can be *varied* only to a relatively small degree, so convective heat loss through the head cannot constitute a major avenue of physiologically thermoregulating thoracic temperature.

The abdomen of most insects, however, is easily insulated from the thorax, and heat flow to it can be physiologically turned on or off. Hence the abdomen can transfer excess heat from the thorax—that is, it can serve as a radiator, or heat sink. Its only drawback is its position (*behind* the thorax), but this is compensated for by its large surface area for heat loss. (The wings are not viable heat radiators because, since they must remain light and unencumbered for flight, only a small amount of heat-carrying hemolymph circulates through them.)

Many large insects from the very diverse orders Lepidoptera, Odonata, Diptera, and Hymenoptera have a fluid-transfer cooling mechanism that dissipates heat through an abdominal radiator while small members of the same groups, or insects that are not strong or continuous flyers, lack the heat-transfer response. Since the cooling function is derived from only small alterations of the already existing system of hemolymph circulation, it can presumably be easily lost or gained through evolutionary time. It seems likely, therefore, that long before the dinosaurs appeared the now extinct giant dragonfly *Meganeura monyi* (see pp. 12 and 18) very likely had, like some present-day dragonflies and possibly all other large, vigorously flying insects, a fluid-cooled flight engine using the abdomen as a heat radiator. At least that would have been the simplest solution, given the available equipment, to an otherwise serious problem.

In insects, as also in automobiles (but unlike in vertebrate animals), the fluid circulating between the site of power and heat production and the heat radiator is without respiratory pigments; it is not involved in gas transfer. In vertebrate animals, of course, a major function of the blood is to transfer respiratory gases, and for that function it contains hemoglo-

In this *Manduca sexta* (dorsal view), the scales have been removed from the top of the abdomen as part of an experiment. The dorsal heart is visible through the transparent cuticle. Note the relatively thin layer of insulating scales on the abdomen and the much thicker layer on the thorax. The curved arrow at the end of the first abdominal segment shows where the heart was tied shut to produce "cardiac arrest" moths that flew but no longer thermoregulated while in continuous flight.

bin to bind and release oxygen. Heat transfer in vertebrates is only an occasional, secondary function of the blood. When we humans are exercising, blood is pumped to the skin or extremities to facilitate heat loss, but this is done at the expense of pumping blood *and oxygen* to the muscles instead. Therefore, work capacity is compromised and distance runners (except apparently those from Kenya and Ethiopia) slow down precipitously in the heat (unless they have another means of cooling). Insect and mechanical flight motors, on the other hand, do not need to compromise aerobic work capacity at higher air temperatures because of thermoregulation. Insects have gas tubes (the tracheae) that reach directly into the muscle cells from small portholes, called "spiracles," along the side of the thorax as well as the abdomen. The total separation of respiratory and heat-transfer functions in insects makes it possible for them to continue working, at least temporarily, even when the fluid flow is interrupted.

Without even doing the experiment, we all know very well what would happen if we attempted to drive a large truck that has its radiator hose pinched shut into the desert in Tucson on a hot summer day. With insects we might be less sure, but in this case equivalent experiments have been performed, by tying off the circulation to the abdomen. These experiments demonstrate that at least sphinx moths, some large dragonflies, and large bumblebees use an abdominal radiator for temperature control of the flight motor. When their "radiator hose" is pinched shut, their cooling response is almost totally abolished and these animals become incapacitated by heat even at moderate air temperatures. Like a Mack truck with a broken radiator hose, they are forced to stop flying as their flight engines approach lethal temperatures.

In insects the "radiator tube" that conducts the hemolymph to the heat radiator also serves as pump, which operates by peristaltic contractions along its entire length. It is generally called the "heart." Although sphinx moths in which surgery has rendered the heart inoperative (by tying it shut) can still fly until reaching near-lethal thoracic temperatures, removal of their insulating layer of thoracic scales makes continuous flight again possible. Since the blood also transfers fuel from depots in the abdomen to the flight motor, however, the duration of flight in animals without heart function is ultimately limited by the availability of fuel supplies stored in the thoracic muscles.

Evaporative Cooling

The mechanisms of increasing or decreasing convection are usually sufficient to stabilize flight-motor temperature near 40–45°C. Evaporative cooling by water necessarily occurs to some extent automatically,

Honeybee workers, *Apis mellifera*, regulate head and thoracic temperature at high ambient temperatures by evaporative cooling from regurgitated liquid (principally dilute nectar).

because all organisms lose water when they exhale and evaporation of water causes cooling. Thus, even though we do not exhale steam on a cold, frosty day to lose heat, evaporative cooling occurs nevertheless. Cooling occurs in this case merely because it is a physical mechanism, not because it is a biological one.

The problem with harnessing evaporative cooling as a mechanism for depressing body temperature is that water is not always conveniently available, and a small animal requires proportionately larger volumes of water to depress body temperature than a large one. Carrying such a large weight would be very costly, in terms of energy, although flight metabolism produces enough water , as a by-product, to suffice for some circumstances. Nevertheless, some insect species, because of their unique feeding habits, do acquire temporary surpluses of water and then use this water for evaporative cooling.

One of the extraordinary examples of an evaporative cooling mechanism in insects is that found in the workers of honeybees, *Apis mellifera,* and yellowjackets, *Vespula* sp. These insects use the head as a radiator, but they do so with a difference. Most radiators retain their fluid, but these radiators are *leaky:* the evaporation of fluid regurgitated from the crop and voided by the mouth causes additional cooling. Because of this system honeybees can fly at higher temperatures than almost any other insect of their size. The disadvantage is that they can't fly for very long, unless there are many flowers with nectar available from which they can replenish their store of water.

Nectar-gathering honeybees normally fly with flight-motor temperatures near 15°C above air temperature. Because of a very efficient counter-current heat exchanger (see pp. 91–92) located between the thorax and abdomen, they are unable to use the abdomen as a heat radiator, as do bumblebees, sphinx moths, and dragonflies. Unlike all other endothermic insects so far investigated, however, honeybees are able to unload *all* of the metabolic heat produced during flight by evaporative cooling; they are capable of the astounding feat of flying even at ambient temperatures near 45°C while maintaining the thorax at the same or only slightly lower temperature.

To reduce the flight-motor temperature excess of some 15°C, honeybees regurgitate nectar from the honeycrop, and while the nectar is held on the mouthparts and the head water evaporates from it. Due to the physical contact between the head and the thorax, thermoregulation of one effectively results in thermoregulation of the other. Thus, the head is cooled by evaporation of water until there is a large temperature difference between the head and the metabolically heated thorax, at which point heat from the thorax follows the temperature gradient and is transmitted to the head.

In honeybees it is head temperature that is actively regulated, with thoracic temperature passively following, since artificial heating of the thorax alone does not result in the heat-dissipation response so long as head temperature remains low. However, artificial heating of the head (as with a narrow beam of light from a heat lamp) almost immediately results in nectar regurgitation and evaporative cooling, even while thoracic temperature is still (momentarily) low.

The honeybees' mechanism for keeping cool may seem almost bizarre, but it could have evolved as only a slight elaboration of their method of making honey, or *vice versa*. To make honey (about 10 percent water) out of nectar (80–90 percent water), bees collect and temporarily store the nectar in their honeycrop, and then they must evaporate off the excess water. They do this in the same way hot bees reduce their body (and hive) temperature, by regurgitating the nectar onto their mouthparts (or the honeycombs in the hive). Thus, honey-making at high temperature accomplishes thermoregulation inadvertently. The bees' mechanism of cooling thus seems almost a matter of elegant inevitability rather than of exquisite design. Yellowjackets, however, do not make honey, yet they have a similar mechanism. Like honeybees, however, they also cool their comb by regurgitating fluid on it.

Where some insects cool evaporatively from the front end, others do so from the back. In the hot Australian desert the larvae of the sawfly, *Perga dorsalis,* in response to solar heat stress first raise their abdomen to the sun to shade the body and to increase convective heat loss. In an emergency, when this response is insufficient, they also

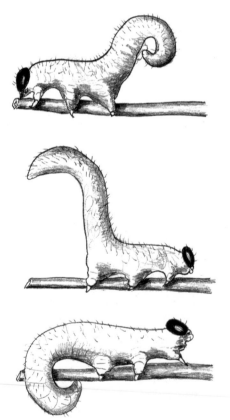

A larva of the Australian sawfly, *Perga dorsalis,* will change position in response to changes in temperature. It is at rest at room temperature *(bottom),* but it raises its abdomen in response to heat stress *(center)* to reduce solar input and promote convective heat loss. At even higher temperatures *(top),* the larva spreads rectal fluid onto its ventral surface to cool evaporatively.

emit rectal fluid and spread it over their ventral surface to cool themselves evaporatively. The response is somewhat analogous to that of storks and vultures who defecate liquid feces onto their legs for evaporative cooling.

Another evaporative cooling mechanism involves the respiratory system. In some desert grasshoppers (presumably well hydrated), opening of the spiracles coupled with abdominal breathing movements to help draw in and to expel air helps increase the volume of air moved through the body. The circulation of air through the body by this "panting" response augments the amount of water loss and causes evaporative cooling, reducing body temperature by 2–3°C below that in animals breathing only to meet respiratory demands. Similarly, the blood-sucking tsetse flies, *Glossina morsitans,* of Africa are often subjected to high heat loads from sunlight while feeding. During a blood meal, however, they experience a temporary excess of water, which they can afford to or must lose for flight economy. They deal both with the excess liquid and the excess heat by opening their spiracles; they cool evaporatively and can continue feeding despite high external heat leaks. Then, after their quick weight loss from evaporation of excess water, they can fly off.

Diceroprocta apache of the Sonoran Desert of the southwestern United States employs a third evaporative cooling mechanism, this one analogous to sweating. These cicadas are plant-sap feeders, and despite living in a dry environment they have access to a large fluid supply by inserting their sucking mouthparts into the xylem of deep-rooted shrubs, such as mesquite. They thus indirectly tap water from deep underground stores. Cicadas sing even when ambient temperatures in the shade reach 40°C, and the repetitive contractions of their tymbal muscles result in internal heat production that adds to the already considerable external heat load. Their singing may raise body temperature 12°C above ambient air temperature.

It had long been a mystery how they could function at such high heat loads, but from recent work by Eric Toolson and associates we now have an answer. Body temperature is reduced to tolerable levels by

The desert cicada, *Diceroprocta apache*, which evades predators by retreating into very high temperatures. It thermoregulates by a response analogous to sweating.

evaporative cooling from fluid shed through large pores distributed over their dorsal body surfaces. The release of this fluid, and the consequent evaporative cooling, occurs only in response to very high body temperature. Most insects, especially those of desert environments, are instead highly resistant to water loss when alive, and upon death there results an immediate *increase* in water loss as the spiracles are no longer actively maintained shut. Killing of the cicada, in contrast, immediately stops the sweating response, therefore showing that it is under metabolic control. The cooling response is mediated, ironically

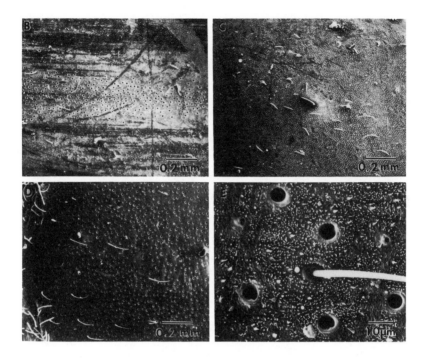

Scanning electron micrographs of the dorsal surface of the desert cicada reveal the pores by which water is lost during evaporative cooling. *(B)* At the mid-line of the thorax the pores are arranged in a line. *(C)* The lateral side of the thorax lacks the pores. *(D)* On the dorsal side of the abdomen, the pores are distributed uniformly. *(E)* In this higher-magnification view both large and small pores are visible.

enough, by aspirin-like substances produced in their bodies in response to heat stress. Because of the sweating response, the male cicadas can sing, and the females search for them, even on the hottest days of the year, when cicada-killing wasps and birds are forced to retire.

Although it may seem counter-intuitive, one may conclude that those insects that heat up the most also have the most effective or elaborate mechanisms of cooling off. Mechanisms of heating up and entirely different mechanisms of active cooling go hand in hand, and both are largely related to body size.

7

Form and Function

Form is an important consideration in the design of all objects, including the bodies of living organisms. Insects are "designed" by evolutionary selection of traits that determine both the mechanical function and appearance of the individual. An insect's shape, for example, will determine how well it flies or crawls. Ornaments and color may be unrelated to mechanical function, but they may be relevant to sexual selection and predator avoidance.

Wing Length and Longevity

The Gossamer Condor, the first airplane propelled by muscle power, can be propelled by man's relatively puny power mainly because it is very light and has large wings. The Condor can't climb and maneuver like an F-16 fighter jet, but it can glide. Similarly, in most insects the relative size of the wings reflects tradeoffs between power (which translates to speed and agility) and energy economy. A bumblebee in flight is a supreme acrobat, but it can glide even less well than an F-16 can, and its flight is energetically more costly per unit weight. However, if fuel is available and speed and maneuverability are at a premium, as when bumblebees are foraging for nectar, then small-winged gas-guzzlers will abound.

The insect gliders include some moths and some butterflies, although neither glide exclusively. They all use their flight motor nearly continuously for gliding wing strokes, or on and off for short glides. Gliding requires large wings relative to body mass, or a low "wing-loading" ratio (body weight/wing area). Low wing-loading can be achieved both by jettisoning abdominal weight and by increasing wing area. In many geometrids (especially winter-flying species such as *Operophtera bruceata* in America and *Erannis defoliara* in Europe), most species of saturniids (including the luna, io, polyphemus, cecropia, and promethea moths), and some sphingids (such as the one-eyed sphinxes, *Smerinthus cerisyi* and *Paonia myops,* and the modest sphinx, *Pachysphinx modesta*), as well as in the mayflies and many other insects from the orders Lepidoptera and Ephemeroptera, abdominal weight is low because the digestive system and/or the egg load has been reduced. Even the hemolymph volume is reduced in some dragonfly species that don't need fluid for cooling a large flight motor.

Adults with low wing-loading that don't feed use fat reserves accumulated during their larval stage to meet their energy needs. They are specialized to be one-time delivery vehicles for a load of sperm or eggs. Perhaps because they have one function for one-time use they need not be durable, and no maintenance is required. Unless they are put in cold storage their adult stage ages rapidly, and it will often die in one or several days.

The high-energy guzzlers are different. They refuel often, make many flights, and live for weeks or months. Examples include bees and sphinx moths, who must stop and start rapidly and maneuver precisely as they visit flowers one after another in rapid succession. To them all of this seemingly frantic activity is worthwhile because nectar is highly concentrated food energy and they get back more energy than they put in. Their very high metabolic rates in turn generate a large temperature excess in the thorax, which places selective pressure to regulate flight-motor temperature so that the individual may be active over a range of ambient temperatures.

Insulation

Many hot-blooded insects that regulate their body temperature have, like their endothermic vertebrate counterparts, bodies wholly or at least partially covered with insulation. Insects have evolved insulation at least three independent times, as we know from the fact that there are three major types of insulation from different phylogenetic origins. One type of insulation, that found in dragonflies, is derived from air sacs. Insects already have air sacs used for breathing, and still air is, next to a vacuum, the best possible insulator. Many insects of various orders have air sacs between the thorax and the abdomen that greatly retard the leakage of heat into the abdomen. But large-bodied dragonflies have gone one step further: their air sacs surround the thoracic flight motor. The other two types of insulation are derived from exterior cuticular structures.

Lepidopterans are covered with a layer of thin overlapping scales that color the wings and body as a mosaic of tiles decorate a parquet floor. In butterflies (and a few moths that fly in the daytime), these colorful tiles are thin and flat and they function in communication; in some species the color patterns produced by the scales are warning signals to predators ("I taste bad" or "I may poison you") or, like racing stripes on hot rods, a sign that says "This model is fast—don't waste your effort trying to outrun it." Other color patterns identify the species and sex; males should not waste their time chasing down inappropriate partners. Most moths fly only at night, however, and the females broadcast alluring perfumes that attract in the dark.

Temperatures are lower at night, and no sun is available for basking. All of this affected the lepidopteran scales: rather than remaining flat and colorful for visual signaling, they have become long and round to form a thick insulating pile or fur coat. In many moths this coating of pile is so effective as insulation it more than halves the rate of heat

This winter moth, *Eupsilia* sp. (Cuculiinae), has thick pile on the thorax for warmth.

loss or doubles the temperature excess, hence permitting flight at much lower air temperatures. Insects with pile now fly in many northern areas, and at times of year where they would otherwise be excluded. Conversely, insects from tropical environments have no or only sparse pile covering.

A covering of setae, small hair-like projections from the cuticle, is the third source of insect insulation. But setae likely have numerous independent evolutionary origins. In many insects small hair-like setae still serve a variety of functions, any one of which could have provided a basis for the modification leading to insulation. For example, the setae of lepidopteran caterpillars are protective: they are spiny and irritating to predators, and even sometimes poison-tipped. In many other insects setae function as tactile organs that detect both sound and other vibrations. In all of these animals the setae necessarily also impede wind movement across the body surface. This would normally have no thermal consequence in caterpillars because they are not endothermic. In some caterpillars that bask, however, the dense packing of the setae reduces convective cooling and helps them maintain an elevated body temperature.

Given an advantage for evolving insulation from setae, presumably by having more of them, there was probably little evolutionary resistance to doing so, because within many other groups, including flies, bees, and beetles, there are some species that are smooth and glabrous while others are heavily insulated with setae. Within the syrphid flies, for example, most species are smooth and bare, but at least in north-temperate regions all over the world a few species are almost as "furry" as bumblebees. Though most bumblebees in turn have very effective insulation, many other species of bees have only a sparse covering of "fur."

Aside from providing insulation, pile or fuzz on the body of a flying insect should also be a source of resistance to air flow. We may wonder how some flyers live in the cold without insulating pile, but it is perhaps more striking that so many insects that fly may still have a layer of pile around the thorax that approximately doubles their cross-sectional profile. It could hardly fail to affect aerodynamic drag, at least at high-enough flight speeds. The evolutionary persistence of pile therefore speaks to its utility.

Two flies that mimic stinging Hymenoptera: the bumblebee-mimicking fly, *Criorhina* sp., with insulating pile *(left)*; the wasp-mimicking fly, *Milesia virginiensis*, without pile *(right)*.

The relative importance of insulation versus flight speed can perhaps best be inferred from examples in bees and wasps. Only the northernmost large bees, the bumblebees, have a heavily insulated flight motor. However, even honeybees have a layer of short insulating pile on the thorax, which aids them on cool mornings and at high elevations (but it is not likely to have any effect on their northern expansion, which is determined by the ability of the *colony* to thermoregulate—see Chapter 13). Bees inhabiting the tropics and hot deserts do not have a covering of pile dense enough to provide appreciable insulation. But even tropical bees cannot get along totally without setae, because they use these projections to trap pollen from flowers. Some wasps, in contrast, live in the same northern areas that bumblebees do, but they do not rely on pollen for protein. Instead, many are predators on fast-flying insects, and they are glabrous. Perhaps an advantage of fuel economy in flight, or perhaps the necessity for fast flight, has in them assumed more importance than drag-inducing insulation for thermoregulatory control.

Color

Color is not trivial, or there would be as many brown cars as red and blue ones on the market. Color is also of vital consequence to insects and other animals: predator avoidance and mate recognition often depend on it, and color also affects the rate of solar heat absorption. Heat from the sun does not reach earth by convection (as heat from a living body dissipates); instead the sun's energy travels in the form of electromagnetic radiation. Light is just the visible portion of the spectrum of radiation, but the visible, infrared, and near-infrared portions could all have a small effect on body temperature. "White" bodies reflect the visible portion of the spectrum, while black represents an absorption of these visible rays. As is true of heat lost by convection, the amount of solar heat absorbed is proportional to body size or relative surface area. In a small body every point in it is close to the surface, and body temperature is strongly affected by heating and cooling events on the surface. On the other hand, the interior of a large body is much more immune to surface phenomena, and is therefore less affected by surface color.

In a few instances an insect's color has a functional significance in thermal balance during basking. For example, lateral-basking sulphur butterflies (*Colias* sp., usually yellow or white) found in cool environments (such as mountaintops or high latitudes) or seasons (early spring) tend to have dark wing-undersides. Ward Watt at Stanford University has determined that these darker individuals are able to heat up the thorax slightly faster than lighter congeners, which buys them additional flight time when basking is needed to prepare for flight. In contrast, the lubber grasshoppers, *Taeniopoda eques,* which

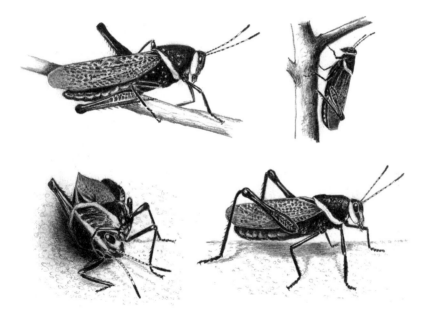

Postures used by lubber grasshoppers, *Taeniopa eques,* to regulate body temperature. The grasshoppers at left are shown flanking to the sun (basking on a twig and on the ground). Those on the right are avoiding heat (perching in the shade and stilting above the hot ground).

are adapted to the *hot* desert environment, are also black (with yellow stripes), although few other grasshoppers are black. In this case, however, the dark color may not only facilitate warm-up by basking in the morning. It may also allow the grasshoppers indirectly to *reduce* their heat load in the afternoon because it gives them new behavioral options.

The black color makes the lubber grasshoppers highly conspicuous and possibly also protects them from ultraviolet radiation. But conspicuousness is an asset rather than a hindrance to them, because they exude a noxious defensive foam that makes them highly distasteful to predators: the black-yellow color pattern is a warning signal. Those predators that have contacted one of these grasshoppers are not likely to soon forget the black-yellow barber-pole pattern. While the conspicuous colors serve as an effective defense, it affords them freedom of movement and opens up a thermal strategy: being protected by their markings, the grasshoppers can safely migrate daily between the hot desert ground and the cooler air of the vegetation above it. That is, they can feed unmolested in the relatively cool air rather than remain hidden on the sun-baked ground as other desert grasshoppers that are not chemically protected are forced to do. Those grasshoppers whose defense is to blend cryptically into the background of sand and debris are forced to endure the sizzling temperatures of up to 50°C that go with the territory.

Although color can have a slight thermal advantage, it is more often subservient to other needs, such as the need to evade predators. Not surprisingly, insects that inhabit open ground often match their background in color and thus they are highly camouflaged. For example, in the tiger beetles, *Neocicindela perhispida,* from the coastal beaches of North Island of New Zealand, one subspecies, *N. p. campbelli,* occurs on black sand beaches and is black in color, while *N. p. giveni,* in contrast, inhabits white quartz sands and is nearly white. Both beetles are camouflaged, but the black beetles pay a price—they absorb more solar heat, and so they lose foraging time because they have to shuttle between sun and shade to keep from overheating. The white subspecies can remain continually active at mid-day.

Stilts, Parasols, and Elytra

A beetle walking on hot desert sand might experience temperatures that could kill it in a minute or less. Just a few millimeters above the ground, however, the hot, ground-hugging air layer is disrupted and mixed with cooler air from above. If we were shrunk to Lilliputian size and forced to live where a few millimeters' difference in elevation could mean the difference between life and death, we would find some way to lift ourselves above the searing heat. Numerous ground-dwelling beetles living on hot sands do just that. Tiger beetles begin to stand tall when sand temperatures exceed 40°C. Aside from extending their jointed legs to stand taller, some beetles, like *Stenocara phalangium* (Tenebrionidae) from the Namib Desert of southern Africa, have evolved very long stilt-like legs that allow them to avoid

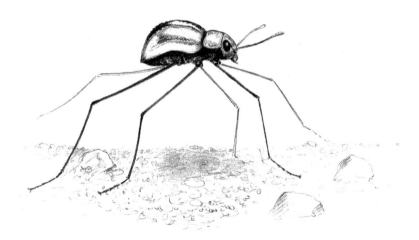

The tenebrionid beetle, *Stenocara phalangium,* from the Namib Desert of southern Africa escapes heat by stilting above the hot ground.

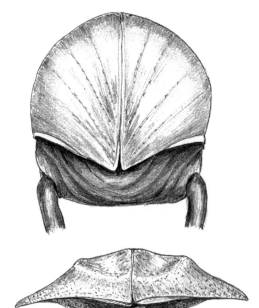

Solar heat input can also be minimized from above. As seen from behind, the tenebrionid beetle at top, *Onymacris unguicalaris,* has domed elytra covering a hollow insulating space over the abdomen. In contrast, the nocturnal tenebrionid, *Stips stali,* is not so protected from direct solar heat input.

overheating by both avoiding the heat at ground level as well as losing some of the solar heat by convection through fast running.

Another option is to use one body part to shade another. For example, like Roman soldiers holding up their shields while marching in the desert, some beetles reduce their absorption of external heat from direct solar radiation by having their own shields (the elytra, which are the modified first set of wings) covering the abdomen. An air space beneath the elytra insulates the abdomen from direct solar radiation.

Rather than having stilts to reduce heat input from below and shields to block solar radiation from above, the larva of the tiger beetle, *Cicindela willistoni,* builds a platform and then perches above the high ground temperatures. In the open, sun-exposed areas of southeastern Arizona where this tiger beetle lives, potential prey also try to escape the hot thermal layer of air near the ground. They perch on the larva's turret to cool off, and the turret then also serves as thermal "bait."

Counter-Currents and Alternate Currents for Heat Retention

The insect flight motor and the sensors and control centers that operate it are located anteriorly, in the head. The storage of the flight fuel is rearward, in the belly. With no passengers except occasional

Like other tiger beetle larvae, the predatory larva of the tiger beetle, *Cicindela willistoni*, builds burrows in the soil. But this species also adds a turret to its tunnel entrance. By waiting for prey at the top of its turret, rather than at ground-level burrow entrances, it escapes the searing ground temperatures and captures other insects that come to perch there.

parasites riding in between the belly and the "cabin," the whole body is awash in fluid that carries dissolved fuel to the flight engines. But mere sloshing and diffusion are not enough to keep the engines running for long. A single large tube, the "heart," runs the length of this airborne marvel of miniaturized design, and this heart has intake valves at the end (near the tip of the abdomen) and along much of its length. Since the heart is a pump as well as a conduit, it forces fluid forward by waves of contractions that travel down its entire length. The insect circulatory system, however, is an "open" system. It lacks veins. Blood leaving the heart percolates freely through the tissues. At the end of each heart contraction, after a pulse of hemolymph enters the thorax and head, the hemolymph seeps through the flight motor, where fuel is removed and heat may be added. The fluid continues on into the abdomen, where a flap of tissue, the ventral diaphragm, undulates like a sheet in the breeze to stir and propel the blood rearward. If the blood is warmer than the air, the heat follows the temperature gradient and is lost to the environment.

At low air temperatures when the abdomen is cool, the flight motor could cool precipitously if the hemolymph carried heat away from the thorax to be dissipated from the abdomen. Two mechanisms, however, normally prevent this potential problem of thoracic cooling. The first is a temporary reduction or elimination of the circulation: if the pumping action of the heart stops, the thorax will remain hot. In an insect, this option is not as far-fetched as it may seem, because insects do not transport respiratory gases in their blood. If we stopped our blood circulation we'd die or be incapacitated in seconds, because all our vital organs would be instantly oxygen-deprived. But in the insect, as in a car, gas exchange would continue exactly as before—the separate system of interconnected air tubes leading directly to the cells from the exterior would be unaffected. Nevertheless, the blood flow to the flight motor cannot be stopped indefinitely because the meager fuel stockpiles that are present in the motor are soon used up.

Another mechanism, one more subtle than cardiac arrest, helps some insects prevent heat leakage from thorax to abdomen. Proof of that is seen in honeybee workers, who *never* show appreciable increases in abdominal temperature, even as the flight motor stays hot. An examination of their circulatory anatomy explains the mystery: they harnessed the principle of counter-current heat transfer long before the technology was engineered by humans.

A *counter*-current is just what its name implies—two currents flowing next to each other but in opposite directions. If the fluid in one current is of a higher temperature than that of the other, then heat (which is not confined by the vessel walls) will passively flow "downhill," from high to low temperature, across these walls. Thus, if the hot blood leaving the thorax flows *around* the vessel in close proximity to cool blood entering it from the abdomen, then heat exchange is inevitable. At least some of the heat from the thorax will be recycled back into the thorax because the incoming blood is heated by the outgoing blood.

To a limited extent the narrow petiole area between thorax and abdomen in *most* insects conforms to a counter-current arrangement. Although this anatomy may in part be almost inevitable, given the overall compartmentalization between thorax and abdomen, these putative counter-current heat exchangers generally exchange very little heat. But minor changes can make a big difference. Counter-current heat exchange can be greatly enhanced merely by prolonging the time or the area, or both, for that potential heat exchange to occur.

In honeybees as well as some other species of bees, both the area for heat exchange and the time needed to accomplish it are enhanced by lengthening the vessel and by winding it into eight tight loops that are crammed into the narrow confines of the petiole area. Before we knew about insect thermoregulation, the function of these loops or coils invited much speculation in the literature. They were variously thought to amplify the pumping of the heart, prevent backflow of blood, or to aerate the blood. We now know that all the blood entering the thorax is cold and that it goes through these loops; all the

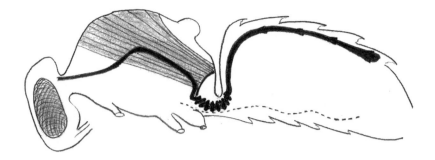

In the honeybee, A. *mellifera* (sagittal section), the convolutions of the aorta in the petiole area increase the time and area for counter-current heat exchange. This ensures that thoracic temperature remains high, which is essential when the bees need to fly at low ambient temperatures.

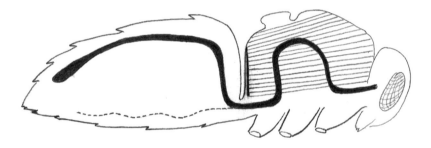

The aortal passage through the petiole in wasps and bumblebees is straight, which allows for more efficient heat transfer to the abdomen, a necessity during brood incubation and flight at higher air temperature.

blood leaving it is at first hot and it is forced to move over or around these coils. The arrangement slows down the blood on its way through the petiole area toward the thorax, and it also increases the surface area for picking up heat. It thus assures counter-current heat exchange and allows honeybees to sequester heat in the thorax instead of having it leak into the abdomen, even as circulation between the thorax and abdomen continues. While the honeybee counter-current heat exchanger functions to prevent heat from leaking out of the thorax to the abdomen, a different arrangement (one found in some other insects) may also be present that reduces heat loss to and through the head.

Heat flow to the head would be reduced if the aorta (the thoracic extension of the abdominal heart) bypasses the hot thoracic muscles entirely. But the aorta loops dorsally through the flight motor (at least in Lepidoptera) in order to bring blood to the wings, which is especially critical as the adult emerges from the pupa, when the soft wings are extended by blood pressure until they dry and harden. Some moths, the Cuculiinae, that fly on cold winter nights tightly loop the vessel back on itself in a hairpin loop that also conforms to a counter-current loop arrangement. The blood gets hotter the farther it flows through the loop and through the thorax, and as a result blood in the descending portion of the loop is warmer than that in the ascending portion coming directly out of the cool abdomen. Some of the heat is recovered from the blood before it enters the head. Additionally, valves in the aorta in the thorax may release the blood *there*, thus reducing the volume of blood flow into the head, and hence the heat loss from it.

Bumblebees and northern vespine wasps have a very much different and seemingly less efficient counter-current heat exchange circulatory anatomy than honeybees and winter moths. This situation may seem counter-intuitive, because they live in colder climates, some species even inhabiting the High Arctic. They might thus be expected to have even better counter-current heat exchangers than honeybees, which are of temperate and tropical origin. But instead of having more loops for counter-current heat exchange, they have none! Nevertheless, their anatomy can also be understood in terms of thermal strategy, but as it relates to their social system. To understand the bumblebees' and the wasps' circulatory anatomy, one must examine their entire life cycle.

Honeybees start their colonies in early summer, when the old queen leaves her colony with tens of thousands of workers in an enveloping

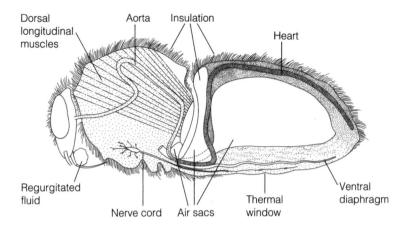

Dorsal longitudinal muscles

Aorta

Insulation

Heart

Regurgitated fluid

Nerve cord

Air sacs

Thermal window

Ventral diaphragm

Structures of the bumblebee anatomy related to heat production and transfer. The dorsal longitudinal muscles and the dorso-ventral muscles (not shown) produce the heat. The hot blood enters the ventrum of the abdomen. Heat from the blood is lost through the ventral abdominal "window," and the cooled blood is pumped back into the thorax by the heart.

swarm. Bumblebee and wasp queens, in contrast, start their colonies very early in the spring, and each new queen attempts this task alone, as an individual. To this individual bee or wasp, time is of the essence, for she must complete the whole colony cycle within a single growing season (the colonies do not survive the winter). A queen's first priority, then, is to rear a group of helpers. Temperatures when and where she builds her nest may be near 0°C, and if the brood were left at that temperature it might take years for them to develop to adults—provided they could withstand the freezing temperature, which they can't. Even in the High Arctic, however, the queens of *Bombus polaris* can produce a batch of workers in about two weeks, as can other bumblebees and *Vespula* wasps. Both bees and wasps accomplish these feats by incubating the brood from the egg to the pupal stage. The queens perch upon their brood clump—consisting of eggs, larvae, and/or pupae—and they press their abdomen upon the brood, much as a hen incubates her eggs with her belly. Only the abdomen provides a smooth surface for contact, but only the thorax produces the heat, by way of intense shivering by the flight muscles. No incubation, and hence social life, would be possible for these insects in a cold northern environment if, like honeybees, they were incapable of transferring heat into the abdomen.

These social insects face a dilemma: they need to transfer heat to the abdomen to heat a small clump of brood and, conversely, they also need to prevent heat flow to the abdomen while foraging at low air temperatures to keep the thorax hot. The solution is an anatomical compromise that permits flexibility through physiological response. The compromise is that the portion of the blood vessel that allows counter-current heat exchange is short and uncoiled, rather than long and looped, relative to the portion found in honeybees. It is long enough to permit moderate heat exchange and hence retention of heat in the thorax, but it is short and straight enough so that a physiological mechanism can be activated that shunts the fluid and heat through quickly, effectively *eliminating* counter-current heat exchange. Furthermore, the transfer of heat through the counter-current heat exchange area is accomplished by a mechanism that is, to my knowledge, not found in any other animal.

In vertebrate animals, counter-current heat exchangers are common, and they are used for heat-conservation. For example, northern marine mammals have them in their flippers, ducks have them in their legs (and so can walk on ice without melting it), beavers use them to prevent heat loss through their tails, and even we have them to conserve heat in our limbs. Counter-current heat exchangers can be by-passed, and heat loss increased, by rerouting the blood into an alternate (generally external) channel. That is why our own veins seem to pop out when we are active in the heat. Such rerouting of the blood from internal to external channels is not possible, however, in insects with "open" circulatory systems lacking veins and capillaries. Instead, at least in bumblebees there is a physiological solution for heat loss in the presence of a counter-current heat exchange anatomy that

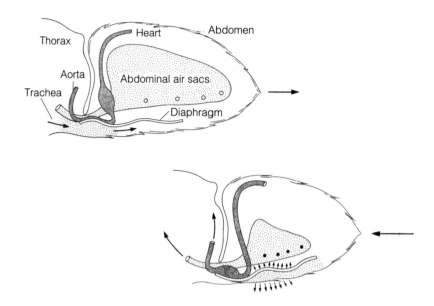

The probable sequence of heart and diaphragm pulsations during the abdominal pumping ("breathing") movements whereby the counter-current heat exchanger in the petiole area is bypassed in *alternate current* heat exchange. By this mechanism bees can heat their brood and incubate their ovaries through the abdomen, with heat generated in the thorax.

serves the same purpose as an alternate blood channel. In the bumblebee this consists of *alternating-current,* or AC, flow of blood. This is the polar opposite of counter-current flow. That is, instead of recovering heat for the thorax, the system acts to remove it, in this case into the abdomen.

Experiments with delicate thermal and mechanical sensors show that to shunt heat past the heat exchanger and into the abdomen, the bee lifts a small valve that allows a pulse of warm blood to enter the abdomen, and in the fraction of a second after the warm blood enters the abdomen, she then squirts a bolus of cool blood into the thorax. And so it goes back and forth, hot and cold pulses of hemolymph alternating in rapid succession. The essential point is that although the blood is not rerouted into a different channel, it is instead temporally "chopped" into alternating pulses in the *same* channels.

The pumping of hemolymph by the heart and the ventral diaphragm is aided also by in-out pumping movements of the whole abdomen, which otherwise function only for moving gas in and out of the thorax; the in-out telescoping movements of the abdomen are synchronous with the heart beats and the ventral diaphragm beats, and they cause pressure changes that facilitate hemolymph flow in precise alternating currents.

We can conclude that in many insects everything from color to covering, from the length of appendages and the proportions of the body to the anatomy of the heart and circulatory system, can and often does have thermoregulatory significance. All of these features are, in effect, both constraints and possibilities for living.

In the following chapters we shall explore the various life strategies in which these thermoregulatory mechanisms play a part.

8

Conserving Energy

One of the primary themes of adaptation—from organisms to ecosystems—is energy economy. The same is true in the design and construction of cars: economy is always a big selling point. Energy saved is energy that can be used for other things. In this regard the insect flight motor is, however, entirely different from a car's engine; the mechanical motor can be made stone-cold dead at *all* temperatures—it consumes fuel only after we turn the ignition. The same is not true for the insect flight motor, for living tissue never totally shuts off; it is always burning fuel, at temperature-dependent rates. Each 10°C rise in body temperature in an organism at rest corresponds to a doubling or tripling of the rate at which energy is used simply to maintain basal metabolism. On an absolute scale the amount of energy used up in this way is small, but if the animal rests for a long time, as many insects do, then the costs can add up, leaving much less energy available for activities such as searching for mates, caring for offspring, or resisting or escaping predators. The fuel required just to stay alive is not a trivial energy drain. Over many days the precious fuel supplies can trickle away. If our cars were like insect flight motors—if they could not be shut off and consumed fuel at temperature-dependent rates—then we would take great pains to park them in the shade or store them on ice when not in use.

By saying that an insect is "at rest," we need not infer the anthropomorphic image of total inactivity. Consider, for example, the sphinx moth *Manduca sexta,* which maintains a thoracic temperature near 40°C during flight and during almost all other activities—all except mating, that is. Copulating moths are decidedly cold-blooded, with body temperature and metabolism strictly at torpor or resting levels. And a single copulation, which can last from 3 to 15 percent of an adult life span, can be a considerable time investment. But since the moths remain poikilothermic (their body temperature matches ambient temperature), this investment during copulation is only about one-fiftieth or one-hundredth what it might be if they remained endothermic. Most (but not all) other insects probably copulate at metabolic rates near torpor as well, especially since in some, such as the love bug, *Plecia nearctica* (Diptera), copulations typically last two days. In *Diapheromea* sp. (Phasmatodea), each copulation (also serving as a form of mate guarding) is reputed to last up to 79 days.

Body temperature may even be kept low when the energy for heating is "free." For example, the adult South African flower scarab, *Pachnoda sinuata,* feeds primarily on flower petals, a rather low-caloric diet of some bulk. During overcast days these beetles munch on petals while perching on top of flowers, but when the sun is out they immediately seek shade under the petals after landing on them. In the shade they maintain body temperatures less than about 30°C, even though they need a thoracic temperature of at least 35°C to lift off to fly. The beetles thus forgo heating by the sun when they can avoid it and will heat up to liftoff temperature (by basking in sunshine, or by shivering if there is overcast) only just prior to attempted takeoff.

The example of the flower scarabs illustrates one of the thermoregulatory options that many insects have that most vertebrate animals lack. Simply by crawling under a flower rather than staying perched on it, a beetle can lower its body temperature by 20°C or more and thus economize in energy expenditure.

The African cetoniid flower beetles, *Pach-noda sinuata* (Scarabeidae), feeding on the anthers, pollen, nectar, and petals of acacia flowers. They situate themselves under the flowers while feeding, but perch on them in the sun for basking prior to take-off.

Energy economy is, of course, particularly critical when a continuously elevated body temperature is maintained by *shivering,* as in bees. Honeybee foragers regulate a thoracic temperature of over 35°C whether they are flying or walking to a food source. When they are kept out of the hive and confined in a small cage, they constantly try to escape and they are continuously flight-ready. They attempt to keep their flight engine warmed up to near 36–40°C at all times, as they need to in order to fly, their usual method of excape. At an air temperature of 23°C, for example, most confined honeybees will die in a mere 5–6 hours because their fuel resources are exhausted. In contrast, Japanese beetles (*Popillia japonica,* Scarabaeidae) of the same mass at the same temperature, which are also capable of shivering to the same flight temperatures, do *not* attempt to stay warm, and they stay alive and healthy for about a week even without food;

their metabolic rate (at 15°C) drops to about one-hundredth their flight metabolism level. This species, which feeds on leaves that yield few calories relative to honey, has evolved to get by on a lower energy budget simply by not shivering for warmth except prior to an occasional short flight.

Energy conservation is even more critical at lower ambient temperatures. If air temperatures are lowered to 10°C, then the hot-blooded honeybees, trying to maintain the same high body temperature (but now at *double* the cost), would live only 2–3 hours in confinement, or half as long as before. The beetles, on the other hand, would live twice as long as before (about 2 weeks): their resting metabolism would be halved rather than doubled because of the 10°C drop in temperature. That is, the beetles would save energy first by choosing not to shiver and second by being in cold storage.

Keeping cool, when it is possible to do so, is the easiest way for insects to reduce energy loss. There are other, more drastic long-term options. These include shutting down metabolic pathways and selective elimination of energy-draining body components. For example, some Arctic caterpillars conserve their energy supplies by eliminating mitochondria when their annual feeding period is past. Colorado potato beetles degenerate their flight muscles during the winter, during hibernation. Milkweed bugs and water striders also degenerate their flight muscles after their dispersal flight, when the flight muscles are no longer needed and instead become an energy drain. They regenerate them later, when flight again becomes useful.

Many winged insects—especially those that cannot periodically atrophy their wing muscles—fly only rarely in their entire lives, and hence for most of their lives they face a problem of keeping cool rather than of staying warm. They only warm up to fly. *Flightless* adults that are seldom subjected to overheating, except for a few instructive exceptions to be discussed, all shun the sun; they have not evolved structures to help capture solar energy because it is important for them to keep cool to conserve energy.

Some insects go to amazing lengths, flying great distances to seek environments where they can keep themselves cool when they cannot feed. Examples include ladybird beetles, Coccinellidae. When aphids (their food) become scarce, these insects migrate up from the hot Central Valley of California to collect in huge masses in the cool mountains for the winter. In the spring, when aphids again become available, they migrate back and resume feeding. Similarly, the monarch butterflies, *Danaus plexippus,* perform spectacular annual flights from all over North America to spend the winter in the cool, high mountains of Mexico. Here they remain in cold storage, maintaining body temperatures of 5–10°C. They arrive each with about 125 mg lipid reserves, on average, and when they leave three months later many have exhausted most of their energy reserves. Had they stayed at lower elevations, at 30°C higher temperatures, they would have depleted their energy supplies at 300–600 percent higher rates, and their food reserves would have been depleted in as little as two weeks.

These energy-saving mechanisms act at the level of behavior to reduce body temperature and thus slow metabolism indirectly, either short- or long-term. There may also be mechanisms of "instantly" slowing down metabolism in insects that normally run fast but make many short stops. Bumblebees come to mind as candidates. Of course, by just "idling," or stopping the shivering activity of the flight motor, these animals save enormous amounts of energy. Nevertheless, in the second that they land on a flower and stop shivering they are *still* hot, near 35°C, and for some seconds or minutes energy drain is still necessarily high. To make another analogy—that the flow of fuel to the mitochondria and their energy mill, the Krebs cycle, is like a running stream—then we could say that the cells have a biochemical pathway (the fructose diphosphate reaction) capable of "backpedalling" against that stream. (In the literature presently available, however, the biochemical backpedalling has been interpreted instead as part of a mechanism of "futile cycling"—making the stream run continuously in the opposite direction—for *consuming* energy in order to produce heat only.)

9

Why Do Insects Thermoregulate?

Throughout this book I've presented examples of the various strategies insects use to survive or take best advantage of the temperatures they encounter in the environment. The individual strategies may be admirable for their ingenuity, but we may wonder if, taken together, they offer any lessons about physiology or evolution. Now that we know something about how insects control body temperature, can we come to any conclusions about why they bother? What is the advantage of thermoregulation?

As it is worded this question cannot be answered, because embedded in it are too many assumptions about evolution. For example, it presumes the unwarranted idea that thermoregulation is a kind of "higher" attribute that has evolved or will evolve in all organisms. Also, it is not clear whether the question should be answered at an immediate or physiological (that is, proximal) level, or an evolutionary (or ultimate) level. The two versions of the answer may be entirely different. Finally, asking the question assumes that there *is* a general answer. More likely, however, thermoregulation has multiple advantages; the mix of selective pressures that has resulted in thermoregulation could vary enormously from one animal to another, though this may not be apparent if one focuses only on vertebrate animals, in which there is a more general uniformity in thermoregulatory physiology.

In spite of these important reservations, I believe enough evidence is availabe for us to take the question seriously. Perhaps one must dare to ask general questions, at least to the level of generality that nature allows.

Advantages for Flight

If we are to consider the advantages of thermoregulation for flight, it is essential that we keep in mind the important distinction between endothermy and thermoregulation. *Endothermy* means, literally, internal heat production. *Thermoregulation* refers to the regulation of body temperature. Although they are commonly assumed to be synonymous, endothermy and thermoregulation are entirely different concepts. The difference may not be very noticeable in vertebrate animals, because those that regulate their body temperature do so to a large extent by producing their own heat (that is, by endothermy). If an insect heats up to a body temperature of 40°C while flying at an ambient temperature of 10°C, however, it is not necessarily thermoregulating; the rise in body temperature may simply be a consequence of endothermy, the heat produced by its flight metabolism. The test of thermoregulation is whether or not the insect maintains a body temperature appropriate for its activity level *independently* of ambient temperature. An insect that must maintain a body temperature of, say, 30°C in order to fly could do so by increasing heat production at *low* ambient temperature, by increasing heat loss at *high* ambient temperature, or by a combination of the above.

We can establish immediately that *endothermy* in insects did not evolve to enhance activity over a range of ambient temperatures. (Thermoregulation might have.) A simple experiment shows this. A tobacco hornworm sphinx moth with its heart tied shut (so it can't thermoregulate) is as endothermic as a healthy moth. Both generate a temperature excess $(T_{tb} - T_a)$ of 22–24°C; in other words, they are highly endothermic. Although both moths are equally endothermic, the non-thermoregulating moth with the tied-off heart can fly only at ambient temperatures of about 15°C–22°C (or at 24°C–30°C, if its thoracic insulation is rubbed off). The unaltered moth, on the other hand, can fly from ambient temperatures of at least 15°C all the way up to near 35°C.

Endothermy, to a lesser or greater degree, is potentially measurable in all flying insects. It is "lesser" in small insects and "higher" in larger ones. The thermoregulation of flyers, on the other hand, is almost exclusively a function of size. Where thermoregulation has evolved in flying insects, it is almost inevitably related to mass; it is not a trait characteristic of phylogenetic development.

Within many insect orders (Hymenoptera, Lepidoptera, Diptera, Coleoptera), mass can vary by nearly a half-million–fold. The largest beetle, for example, may be 350,000 times heavier than the smallest: beetles of the Ptiliidae weigh considerably less than 1 *milligram*, while the Goliath beetle, *Megasoma elephas* (Scarabaeidae), weighs up to 35 *grams*. The largest beetles necessarily heat up in flight, whereas the small ones do not. Medium-sized beetles, such as the 100-mg Japanese beetles, *P. japonica*, heat up only 2–3°C in flight, but the large beetles (those weighing more than a gram) heat up to above 40°C even at modest air temperatures and cannot allow themselves to heat up much further.

Large insects thermoregulate for two reasons. Proximally, they require means of heating up *prior* to flight and they need to limit their temperature excess once in flight (this is the definition of thermoregulation). A small beetle, on the other hand, does not overheat easily even at high ambient temperatures, and it can negotiate over a wide temperature range without thermoregulation. Thus, large, strong flyers must thermoregulate to be active over a range of air temperatures, or else they would be restricted to flying in a narrow range of low air temperatures they may find in some suitable habitat. But what about an insect that engages in both flight locomotion (where it is necessarily endothermic) and terrestrial locomotion (which generates little body heat for insects)?

Once having evolved the capacity (out of necessity) to operate the thoracic muscles at a high temperature for flight, insects may have the option to stay hot while walking or running. For example, many beetles that *fly* may also show considerable endothermy when they run, but no non-flying beetle is endothermic when it runs. However, a beetle that thermoregulates during running or walking on the ground may require thermoregulation for something entirely different than the simple need to continue locomotion. The reason might be, for example, to gain an advantage during competition over food or mates. A beetle could potentially achieve the same high body temperature by either basking or shivering. But if no *small* beetle heats up for contest competition on the ground by either mechanism, and if only those large ones that can fly do so, then we might conclude that the ultimate reason the large, flying beetles thermoregulate during terrestrial activity is that their size requires them to do so during flight. Thermoregulation could now also be seen as being related to competition on the ground, because it is proximally expressed in direct response to competitors for food or because it has evolved to serve that function and is expressed at all times. In both cases the behavior could be genetically coded.

Although thoracic temperature is in part dependent on body size, head and abdominal temperatures need not be size-related. To some degree, therefore, when thermoregulation of the latter body parts occurs, it may be more directly and specifically related to function. For example, the large dragonfly, *Anax junius,* is a hunter that relies on vision and very fast reaction times to capture prey, and in this species head (and eye) temperature is elevated by passive conduction and by physiologically facilitated heat transfer from the thorax at low ambient temperature. The visual physiology for prey location is likely dependent on appropriate eye temperature, and the regulation of head temperature probably serves to regulate eye temperature.

As another scenario for the "why" of thermoregulation, consider a large, endothermic flyer that evolved to become a sit-and-wait strategist rather than a continuous flyer. Niko Tinbergen and associates published in 1942 a study showing how grayling butterfly males, *Eumenis* (now *Hipparchia*) *semele,* defend patches of open sand that serve as mating territories. They dangled model butterflies from fishing poles near these males, and the males rose from the sand, where they had been perched, and approached or pursued some of the models. Normally the butterflies make short, dashing flights to intercept passing females, or to do aerial combat with passing males and chase them away. In order to perform these two functions, perching males must be instantly flight-ready. For that they need to regulate their thoracic (and head?) temperature, and that has its costs.

As in some other Satyridae, grayling butterflies bask by tilting their wings (and whole body) perpendicular to the direction of the sun's rays, thereby maximizing the body area exposed to the sunshine. This behavior may be costly in time, and it may compromise the animal's vision in the one eye pointed downward. As in other butterflies, the duration of perching bouts by graylings increases greatly at lower temperatures. Might the necessity of having to maintain a high thoracic temperature (in other words, to thermoregulate) not be disadvantageous because it takes time and compromises vigilance?

Many butterflies may fly nearly continuously at high ambient temperatures, when the optimal thoracic temperature is achieved automatically with little or no thermoregulation. At lower air temperatures, however, the thoracic temperature achieved during flight is too low, and the butterflies must then stop frequently and often bask for long durations. Would it not be better not to have to take the time to bask (thermoregulate) but be able to fly longer at lower air temperatures also? The answer might depend on body size.

Both body mass and air temperature determine an insect's activity regimen, since the lower the air temperature and the smaller the body mass, the shorter the flight durations and the longer the periods of basking must become to keep a suitable thoracic temperature. In this case, thermoregulation is immediately or proximally advantageous and necessary for, say, mate- and prey-catching and possibly for other functions. But thermoregulation is ultimately a *dis*advantage if it is seen in terms of the costs and the alternative of evolving the ability to operate the flight muscles at a lower temperature. Thermoregulation might just be the simplest available option that evolution would use as a solution if environmental temperatures are consistently lower and/or body mass is consistently less. That is, thermoregulation is proximally a necessity and hence an advantage. *Ultimately,* however, it may be more advantageous to adjust the muscle physiology so that the insect could fly at a lower thoracic temperature—in other words, dispense with thermoregulation so that continuous flight is possible without thermoregulatory behavior. As discussed previously, however, biophysical limits dictate the amount of endothermy, and muscle physiology is evolutionarily constrained by these biophysical dictates.

Advantages for Growth

Most insect larvae are soft-bodied, highly nutritious, and sluggish, making them prized food items for a variety of predators, notably birds. Larvae have no significant muscle mass for internal heat production, and hence they have limited options for raising their body temperature except by basking. But basking puts them in a dilemma: an elevated body temperature greatly accelerates feeding, digestion, and hence growth rate, thereby reducing the time they spend as easy prey, but any advantage they gain of reduced predation at a future time (by reaching the more mobile adult stage earlier) is bought with an increase in exposure to predators in the present. In addition, parasites may also take their toll during the vulnerable larval stage. Under temperature extremes, however, basking may not be an option. It could be a necessity.

In the High Arctic the caterpillars of the *Gynaephora groenlandica* moth orient almost continuously to sunshine during their brief period of summer activity. Basking caterpillars are several degrees above the ambient temperature, which is usually less than 10°C. But even with basking, the *Gynaephora* caterpillars living near the far northern

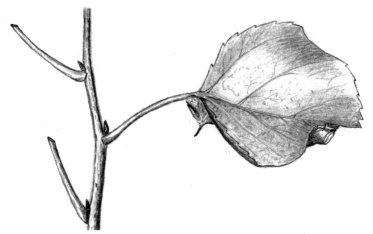

Three cryptic caterpillars that are highly prized by birds. Their predator-avoidance behaviors make thermoregulatory behavior impossible. *Above:* Modest or poplar sphinx, *Pachysphinx modesta,* feeds in the daytime but remains under leaves. *Below left:* Underwing moth, *Catocola* sp., appressed to a poplar twig. Note head of caterpillar at lower end and fake "shadow band" across body at upper end. *Below right:* A large geometrid caterpillar mimicking a lateral twig. Neither of the latter species move at all all day long.

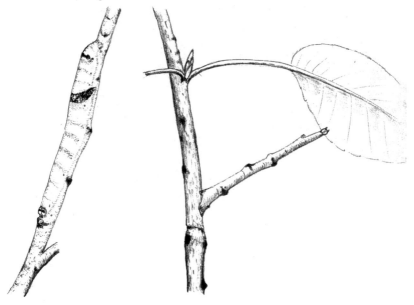

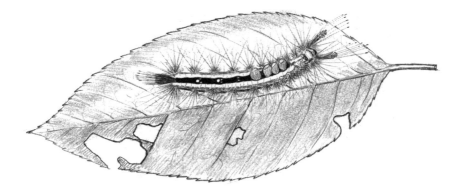

A tussock moth caterpillar, *Hemerocampus leucostigma*, is colorful and equipped with highly ornamental but noxious tufts of "hair." This caterpillar is distasteful to birds, and it exposes itself on leaf surfaces in sunshine.

The caterpillar of the Arctic moth, *Gynophora groenlandica*, is thickly furred. The caterpillar stays in the open and basks nearly continuously day and night (where there is midnight sun). The fur reduces convective heat loss. Here it is shown in a typical habitat, at the edge of the melting ice at Ellesmere Island in the Canadian Arctic.

The cocoons of *G. groenlandica* also bask. Attached to a rock, a cocoon appears to be oriented to maximize exposure to sunshine and reduce convective cooling. It may also serve as a miniature greenhouse to heat the occupant.

boundary of their range still require about thirteen years to grow from egg to pupa, in part because the presence of tachinid fly and ichneumonid wasp parasites limits the already very short period (the season when snow-free ground and sunshine are available) when the caterpillars may bask. That is, soon after they emerge in the summer, the caterpillars must again hide—they literally "go on ice" on the permafrost until the next summer.

The High Arctic *Gynaephora* pupae complete their development to adulthood in only one summer, probably because the caterpillars, unlike all others, spin their cocoons on top of rocks and other exposed places where they can be warmed by the sun twenty-four hours per day. It is safe to conclude that if both the caterpillars and pupae did not bask, then development to adulthood would be delayed by decades and/or their range would not extend to the High Arctic.

For some grasshoppers living in open rangeland in the western United States, basking permits the animals to complete their life cycle within one season. Modeling calculations, by Raymond Carruthers and colleagues from the U.S. Department of Agriculture, based on temperature and growth rate indicate that without basking the animals could not complete their life cycle in the one year that seems to be allotted to these species in their present evolutionarily constrained life cycle. In social insects (see Chapter 13), thermoregulation of the larvae is accomplished by the adults, who create near-tropical conditions in the nest so that the larvae reach adulthood in about two weeks.

Advantages for Egg Incubation

The sun shines at midnight on Ellesmere Island in the Canadian High Arctic, but even so ambient temperatures barely exceed 5°C. In mid-June the purple saxifrage and creeping arctic willow start to bloom on bare patches of ground that the ice has just left. Insects seem virtually absent from the flowers, except for an occasional queen of the large wooly bumblebee, *Bombus polaris*. This species has the distinction of being the northernmost social insect in the world. These gravid bumblebee females are the sole winter survivors of the population, and they would now mature the eggs in their ovaries, start nests, and finish their colony cycles before another six weeks. The essential key to their remarkable performance is thermoregulation, which provides the larvae with a near tropical environment in which

A *Bombus polaris* queen looks like a giant alongside the erect catkins of Arctic willow. In these arctic bees gravid females maintain an abdominal temperature of 30°C or more, probably to promote rapid egg development.

to grow. Thermoregulation may also accelerate egg development in the ovaries, to make it possible to squeeze the social cycle into the short growing season.

In most insects, as we have seen, the temperature of the abdomen remains low, passively following that of the external environment, or it heats up secondarily if excess heat must be dissipated from the thorax. However, the queens of *Bombus polaris* and other Arctic species that forage in early spring far above the Arctic Circle have conspicuously *elevated* and stable abdominal temperatures, while queens of other species from the northeastern United States have lower and more variable abdominal temperatures, as most other insects do. Might the gravid females (queens) from the Arctic be incubating their ovaries? It appears so, because the sterile workers and the drones of *B. polaris* have lower abdominal temperatures (at given ambient temperatures) than the fertile queens, and their temperatures are indistinguishable from those of temperate bumblebees. Furthermore, in one experiment F. Daniel Vogt and colleagues found that bumblebee ovary mass increased from 1 percent of body mass to 3 percent of body mass in 6 days after emergence from hibernation if abdominal temperature was maintained at 32°C, whereas there was almost no change in ovary mass if abdominal temperature remained near 24°C. Thus, although the abdominal temperatures would normally remain low, perhaps near 10–15°C, because of the external environment, the abdominal temperature that the Arctic queen bumblebees maintain (over 30°C) can be considered "incubation" of their ovaries. The increased abdominal temperature speeds up the rate of egg production in the same way that later incubation of the eggs and larvae in the nest speeds up growth for the rest of the bumblebee developmental cycle.

A similar situation is found in reptiles, in whom viviparity (bearing of live young) is common in colder habitats. Pregnant females keep warm by basking, and as a consequence the embryos grow faster inside the body than outside. Warming by internal incubation in lizards can accelerate development by a month, and the strategy is especially useful where summers are short and the eggs need to hatch before the onset of winter.

Advantages of Avoiding Temperature Extremes

Animals can evolve to tolerate any of a wide range of temperatures, and thermoregulation is often not required for a species to survive. Thermoregulation is essential, however, when large potential differences in body temperature are dictated by the physical environment over a short time frame. Large differences may occur not only in terms of rest versus flight exercise but also with respect to environmental factors. An excellent example of the latter can be seen in a recently published study by Brent A. Ybarrondo of Adams State College in Colorado. Ybarrondo compared habitat selection and thermal preferences in two species of water-scavenger beetles (Hydrophilidae). One species, *Tropisternus mixtus,* inhabits cool ponds in Vermont, and the other, *T. columbianus,* inhabits thermal pools in Yellowstone National Park, where small pools separated by less than 1 meter may differ by as much as 75°C. The beetles disperse by flying from pool to pool, a potentially dangerous behavior for *T. columbianus.*

There are no serious thermal consequences for a Vermont beetle staying in one pool versus another, because at any one time temperatures in all pools are roughly equivalent and far below the beetles' thermal death point of 42–43°C. A beetle in Yellowstone Park who lands in a pool of the wrong temperature, however, may have to react quickly or die. Indeed, Ybarrondo found that the beetles sometimes make the wrong choices and are killed; he found dead beetles at a Yellowstone pool of 73°C. Presumably, beetles landing in such pools are heat-killed before they have a chance to get out. But at most hot pools they can, and do, react quickly.

To examine the beetles' responses, Ybarrondo compared the two species in the laboratory at the University of Vermont. He released beetles that had been first equilibrated to either 25°C or 35°C in a linear tank that maintained a temperature gradient from 10°C to 50°C with a cooling unit at one end and a heater at the other. He found that the *T. mixtus* (the beetles from the thermally uniform and cool environment) were poor thermoregulators. Almost 20 percent of them wandered into the hot zone and were killed, although after a half-hour to an hour most of them eventually settled in an area near 27°C. In contrast, the Yellowstone beetles also settled at near 27°C but they did so within only 5–10 minutes. In other words, they chose the same "comfort" temperature in the temperature gradient but they reacted much more rapidly. Almost none were heat-killed, because they quickly turned around and avoided the hot end of the temperature gradient. Ybarrondo postulates that the "hasty retreat" made by the Yellowstone beetles is a response that evolved in *T. columbianus* because this species experienced a strong selective pressure for the ability to detect and respond quickly to extreme variability in temperature. That is, they evolved to become excellent behavioral thermoregulators because of the great temperature differences in their environment.

Alternatives to Thermoregulation

Many insects heat up only because of their flight metabolism, and if they are small they have little chance of ever heating up, much less of thermoregulating. Nevertheless, insects such as gnats, fleas, flea beetles, springtails, and leafhoppers may be active at very low temperatures because their biochemical machinery has adapted to operate at the temperatures they experience, as of course large insects do too. For example, the adults of the Himalayan glacier midge (*Diamesa* sp.) walk on snow surfaces as cold as −16°C. In New England, winter gnats, *Trichocera saltater,* may fly at air (and body) temperatures near 0°C, and almost microscopic "no-see-ums" (midges) fly only at *high* ambient temperatures in the summer. It is doubtful that the no-see-um could fly at low temperatures and the gnats at high (*Diamesa* midges are killed by the heat of ones' hand). However, some insects have succeeded in side-stepping thermoregulation and finding a means of

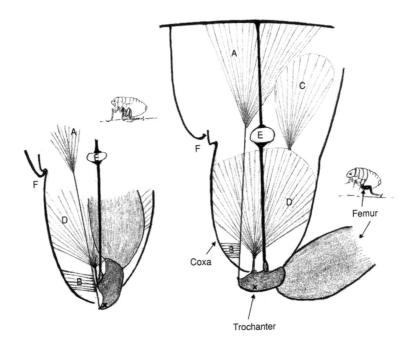

The jumping mechanism of the flea. Fleas are flightless insects derived from a winged forebear in whom some of the flight muscles have been incorporated into the jump mechanisms. At left a flea is shown in the "cocked" position; the energy for the jump is already stored (in the contracted muscles A and D and in the rubber-like resilin pad, E) and ready to be released. Pulling B, the "trigger" muscle, puts the forces off-center of the pivot point (x) on the trochanter, and the stored energies are released.

locomotion that works not at one but over a wide range of ambient temperatures. They do it by jumping.

Small mass is especially advantageous for jumping, and fleas (Siphonaptera) are of course famous for it. But a variety of leafhoppers (Homoptera), springtails (Collembola), and flea beetles (Coleoptera), also jump prodigious distances—sometimes hundreds of times their body length. They accomplish these feats by not by thermoregulating (that is, by maintaining a body temperature high enough to allow more rapid and stronger muscle twitches) but by storing energy prior to the jump and then releasing it instantaneously. In all of the different jumping mechanisms, prior to the leap there is a slow muscle contraction, analogous to pulling the string of a bow, and this stage of energy storage is temperature-dependent. But the release of the stored energy is independent of temperature. (The same type of trip or cocked-spring mechanism has recently also been found in the fast "trap jaw" mechanism of a ponerine ant. The ant uses it not for jumping, however, but for snagging its mandibles onto prey).

Jumping is not the only alternative for locomotion available to small insects that do not thermoregulate. The small, winter-flying moths of

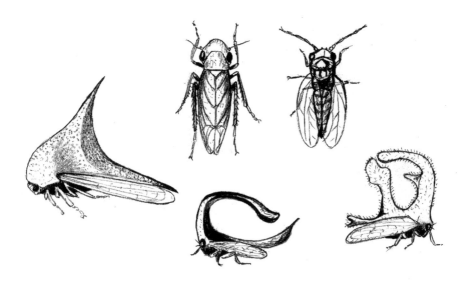

Some of the jumping homopterans. These insects, like fleas, collembolans, grasshoppers, flea beetles, click beetles, and others, jump prodigious distances even without having hot, and fast, muscles. Energy for the jump is stored by a slow muscle contraction analogous to pulling a bow, and all of the stored energy is later released, much as one releases an arrow to shoot. Clockwise from top center: a leaf-hopper, *Tetigella* sp. (Cicadellidae); a plant louse, *Psylla* sp. (Psyllidae); and three treehoppers (Membracidae).

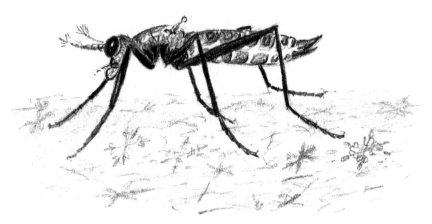

The flightless crane fly, *Chionea vulga* (Diptera), walking on snow. Like many insects that are active in places or seasons too cold for them to generate sufficient power for lift-off for the flight muscles, they lose their wings. The dictum to "use it or lose it" seems to apply literally.

the species *Operophtera bruceata* commonly flutter both in the daytime and at night in the New England woodlands. The males are much too small (mass is about 9–12 mg) to keep warm by endothermy, but they are able to fly at an extraordinary low body temperature near 0°C. This is possible for several reasons: alterations of muscle physiology, their extremely small body size (their digestive tract has been jettisoned, as the adults do not feed), and the very much enlarged wings (which act in part as sails). The females, which also don't feed as adults, are already fully loaded with eggs upon emergence from their pupae. Living at the low-temperature limit, they would be unable to lift off because they would be unable to generate sufficient power when heavily loaded with eggs. Wings are thus useless for the females, and the muscles to power them have been eliminated through evolution. Having given up flight entirely, weight no longer matters, and females have evolved a very large size (about four times that of males). This weight has been allocated primarily for eggs. If females were to fly over the same range of muscle temperature as the males, they could carry only a very few eggs and their fecundity would be greatly reduced. In these moths, and in many other winter-active insects that follow a similar thermal strategy, energy and other food resources are harvested in the larval stage and in adulthood one sex is specialized for gene dispersal. The other sex can then relinquish locomotion and the attendant costs of thermoregulation to achieve it. That is, the otherwise enormous costs of thermoregulation have been eliminated by adopting a life strategy that requires no thermoregulation at all! When activity *is* shifted to still lower temperatures, as in some winter-active flies, then even the males may lose their wings.

10

Strategies for Survival

Hot Harvesters, Cold Sleepers

A Japanese beetle could satisfy its food requirements for a week or more by remaining in place on a single large leaf and feeding on it. A vastly different situation confronts nectar foragers. Flowers provide sufficiently concentrated food to attract pollinators from afar, but this food is commonly provided in almost microscopic packets that keep the pollinator moving from flower to flower. In general, a bee extracts the energy resources from a flower in a second or less, and it must then fly on to the next flower. It needs a large flight motor for acrobatic flight, and under most foraging situations it must keep its thoracic muscles hot virtually continuously. Most nectar-foragers therefore have no choice but to invest heavily in thermoregulation and flight metabolism to ensure paybacks; at most kinds of flowers no energy profits can be made unless large energy investments are almost continuously made in thermoregulation.

In bumblebees, the energy costs of foraging include the energy used for flight and for shivering (at low air temperatures) during the seconds or fractions of seconds while the bee is perched on flowers.

In order for a bumblebee to forage at 10°C, for example, it must maintain the temperature of its flight motor at least 22°C above air temperature if it is to be instantly flight-ready for successive flower visits. The bee must therefore shiver vigorously in the inter-flight intervals, when it would otherwise cool precipitously. The energetic cost of keeping the flight motor heated can be as high as the cost of flight itself. The bees roughly balance costs against profits, however, in that they avoid foraging from flowers that give low reward (where the combined costs of flight and thermoregulation are near or in excess of potential profits).

As a means of maintaining the "profit margin," body temperature varies as a function of expected profits. When food rewards are very rich, then bumblebees and honeybees expend more energy for flight and shivering, and they also maintain higher flight-motor temperatures while foraging. Honeybees will even maintain higher temperatures while dancing to advertise the location of richer food resources being harvested. Higher thoracic temperatures permit faster foraging speeds and profits per unit time increase. Energy investment during foraging is based not on stomach contents that directly fuel energy expenditure but on the *expectation* of rewards. For example, when only microscopic nectar droplets are available but they are distributed in clusters (as on large inflorescences, where short inter-flower flights and very long perching durations are possible), then body temperature may decline even though large volumes of fuel may ultimately accumulate in their fuel tank, the honeycrop.

Even insects that maintain high temperatures for some activities, such as foraging, may at other times spend long periods at much lower body temperatures. The intense thermoregulatory phase of the queens of both bumblebees and wasps, for example, is short relative to their entire life cycle. In the northern part of their range, the queens of wasps and bumblebees may spend most of their lives in a state of suspended animation, where their body temperature is at or below the freezing point of water. We know relatively little about this part of their lives, since the animals are hidden in the ground, and few people have ever found the hibernating queen, and fewer still have been prepared to know what to look for.

It was therefore a treat for me to talk recently with my friend and colleague Brian Barns at the University of Alaska at Fairbanks. Brian has intensively studied hibernation in Arctic ground squirrels, which has prepared him to look at hibernation in other animals, such as grizzly bears, bees, and wasps. In the summer of 1994 there was an unusually large outbreak of common yellowjackets, *Vespula vulgaris* and *V. arenaria,* near Fairbanks, which encouraged him to try to find the overwintering queens in hibernation. He and his students dug in the leaf litter, and they were lucky to have looked in the right type of hibernating area for *V. vulgaris,* because they found dozens of them (but none of *V. arenaria* nor bumblebees that also hibernate underground).

The yellowjackets had not merely burrowed into the soil. Some had excavated little chambers or pits, within which they hung suspended by their mouthparts! Sleeping all winter suspended by the jaws is an odd behavior. It could be significant. And it was.

The key to understanding this discovery is likely to be "supercooling," the phenomenon of a fluid remaining unfrozen even though its temperature is below the freezing point. Supercooled liquids are in a physically unstable state and will freeze instantly if brought into contact with substances, such as ice crystals, upon which other ice crystals can form. In late September, when he dug up the first queens, Barnes found that they did not freeze until near −7°C, but when they did start to freeze they turned to ice instantly. And that is where the hanging-by-the-mouthparts behavior is probably significant.

Although some insects (and a few amphibians and reptiles) have evolved the amazing capacity to survive having their extracellular water frozen into ice, most avoid this route of suspended animation and instead survive by supercooling. (The hazard of this strategy is that if supercooled insects *do* freeze, they are invariably killed because there is not enough time to adjust physiologically and lethal *intra*cellular freezing occurs.) Supercooling can occur only in the absence of ice crystals. That is, a supercooled animal freezes almost instantly, in a process known as "flashing," when even one ice crystal comes in contact with its body fluids, because supercooling occurs only when there are no templates on which ice crystals may form. Normally the best template is ice itself. If insects are buried in moist soil or wet leaves, the water surrounding them becomes in winter a bed of fatal ice crystals. But by hanging in a little cavern, or in layers of dry leaves away from draining water, the insect effectively removes itself from the external offending crystals and therefore has a chance to supercool.

Exposure to ice crystals can also be minimized by hibernating in a dry place (though one must then ask if there are dry places in the tundra, these insects' usual habitat). Doing so could be dangerous, however, as it could result in the animal dying of dehydration during the nearly 9–10 months that the northern wasp and bumblebee queens remain buried and dormant. Hanging in a moist place might be the ideal solution by which the queen could have it both ways—remain moist and not freeze—for in this way the insect would be surrounded by air, not water, and less likely to encounter ice.

*Un*frozen animals remain responsive—they are open to appropriate environmental cues in the spring and can perhaps emerge earlier from the ground to begin their colony cycles. The downside of being unfrozen is that, no matter how cold the animals are, their metabolism is never totally shut off, so they have to live for many months without feeding. But they prepare for their long fast. Barnes noted that in the fall the wasps' abdomens were "loaded with pure white dense pasty material" that was either fat or glycogen.

The ability to supercool appeared to be at least to some extent facilitated over time by a physiological response. Late-winter wasps remained unfrozen and supercooled at near −15°C, whereas near the beginning of hibernation they had remained unfrozen down to only −7°C. Temperatures in leaf litter where the queens hibernated under 2–3 feet of snow never dropped below −6°C all winter, even at air temperatures of −40°C. Hence, the wasps' supercooling ability affords them complete safety (provided they avoid contact with ice), and it guarantees them a very low energy expenditure to help make their reserves last 9–10 months.

Some insects (and some polar fish) contain proteins in the blood that bind to ice crystals in such a way that they prevent the formation of more such crystals in the blood. Whether this occurs in the wasps is anybody's guess, but the suite of observations again shows the important role body temperature plays in survival in a hostile environment, where there is a time to be hot, and usually a (much longer) time to be cold.

Hot-Bodied Warriors

Body temperature may not only be varied as a response to environmental cues, such as the availability of food resources, it may also be varied in response to other insects. Fighting—to take resources from others—is one example. Contest competition or fighting over food is rare in insects, but at least two species of African dung beetles, *Scarabaeus laevistriatus* and *Kheper nigroaeneus,* engage in combat over dung balls that they make both to feed on and/or to serve as sexual attractants. An elevated thoracic temperature plays a crucial role in these contests. In contrast, those beetles that do not engage in contests are hot only in flight.

A pair of dungball-roller beetles. The males of some species shiver to keep their flight and leg muscles warm, thereby increasing ball-rolling velocity. The female, who rides passively on the ball, cools down after attaching to the ball.

Both species make dung balls, up to tennis-ball size, from the droppings of a variety of African plains animals, ranging from antelopes to primates to elephants. Like other dung-ball rollers, the beetles roll their freshly made balls away from the pile of droppings and bury them. Once buried, the balls are used, in relative safety, as food either for themselves or for a mate and larvae. Making and rolling the balls above ground, however, exposes the beetles to the risk of predation by birds and mammals. Many beetles crowd their activity into the night hours, when they can avoid predatory hornbills and other birds, but they then face intense "scramble competition" from other dung beetles following the same strategy. On one research trip to Tsavo Park in Kenya, George Bartholomew and I counted thousands of beetles that came one night to a single elephant dropping in fifteen minutes. Most of these were tiny, poikilothermic individuals that burrowed into the dungpile to feed there. But the beetle horde also included large rollers that tried to remove the dung from competing beetles to feed on it in isolation.

Combat over already-made dung balls by the nocturnal dungball roller, *Scarabaeus laevistriatus*. Hot beetles defeat those with lower thoracic temperature and expropriate their balls.

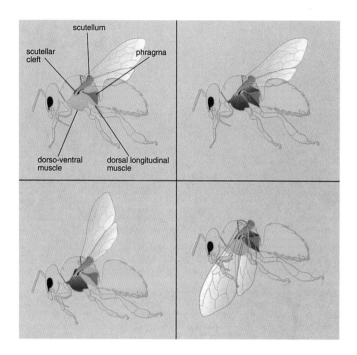

Warm-up proceeds without visible movement in some bees. The process involves the flight muscles (pink)—the dorsal longitudinal and dorso-ventral muscles, which move the wings indirectly through levers—and the scutellum (the rear portion of the thorax, which projects downward as the scutellar arms). The dorso-ventral muscles extend from the top to the bottom of the thorax; the dorsal longitudinal muscles extend from the top of the thorax to a freely moving structure called the phragma, which pushes the scutellar arm. Before warm-up *(top left)*, the flight muscles are relaxed, leaving an opening, the scutellar cleft, between a scutellar arm and the front portion of the thorax. During warm-up *(top right)*, both sets of flight muscles contract simultaneously (red), but the dorsal longitudinal muscles contract more. This imbalance of contraction rotates the scutellar arms into the thorax, creating a mechanical stop that prevents further movement, thereby generating motionless warm-up. Flight begins with a sudden contraction of the dorso-ventral muscles, which rotates the scutellar arms away from the mechanical stop and raises the wings *(bottom left)*. The dorsal longitudinal muscles then contract, depressing the wings *(bottom right)*. The upstroke muscles contract again before the scutellar arms hit the mechanical stop. (From B. Heinrich and Harald Esch, "Thermoreg-ulation in bees," *American Scientist* 82 (March–April 1994), 164–170; reproduced with permission.)

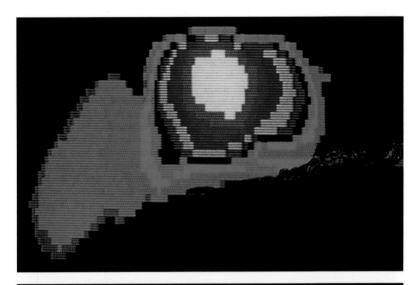

Thermograms of *(top)* a flying honeybee, *Apis mellifera,* and *(bottom)* a shivering drone fly, *Eristalis tenax,* which mimics the honeybee. For both insects, the highest temperatures (yellow) are found at the center of the thorax, and the lowest temperatures (blue) are furthest from the thorax. In the honeybee there is heat transfer between thorax and head, but not between thorax and abdomen. The fly, in contrast, exhibits no heat transfer: both head and abdomen are cooler than the thorax. (From B. Heinrich, "Insect thermoregulation," *Endeavour* 19 (1995), 28–33; reproduced with permission.)

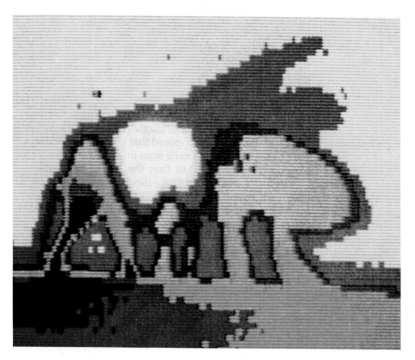

A honeybee worker imbibing sugar-water must shiver to keep warm. (The sugar-water is shown here in black, which indicates the lowest temperature.) The thorax is maintained at the highest temperature (white) by the bee's shivering, but the abdomen stays cool (green). (Courtesy of Sigurd Schmaranzer and Anton Stabentheiner. From B. Heinrich and Harald Esch, "Thermoregulation in bees," *American Scientist* 82 (March–April 1994), 164–170; reproduced with permission.)

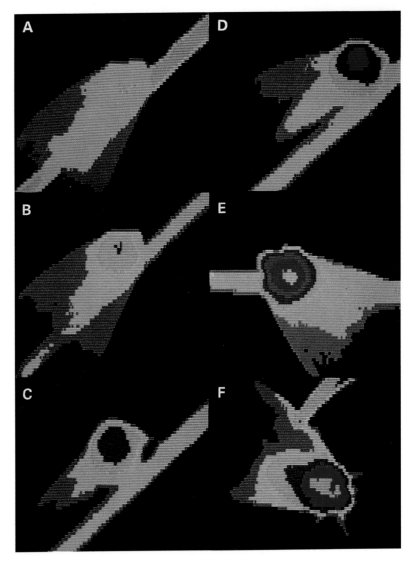

Infrared thermovision photographs of a winter moth, *Eupsilia morrisoni* (length = 1.7 cm), against a background of ice (black). During pre-flight warm-up (*A–E*), the moth gradually increases its thoracic temperature (as indicated by the change from light blue to yellow). After one minute of flight (*F*, photographed while the moth is held with forceps by its left wing), the highest temperature of 26.6–30.9°C is measured. Note that no heat is transferred to head, abdomen, or wings. (From B. Heinrich, "Thermoregulation by winter-flying endothermic moths," *Journal of Experimental Biology* 127 (1987), 313–332.)

Two winter moths *(Eupsilia vinulenta* and *Orthosia rubescens)* blending in with the leaves where they hide in the winter.

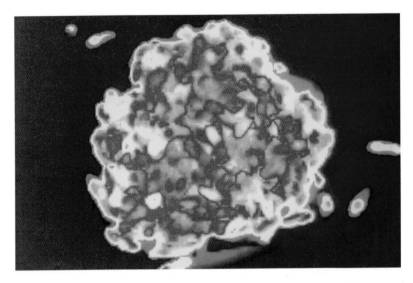

Color thermograph showing a hot defensive ball of *Apis cerana japonica*. (Courtesy of Masato Ono.)

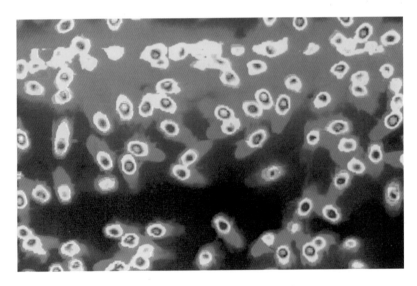

Color thermograph showing defender bees crawling around nest entrance. (Courtesy of Masato Ono.)

Defensive ball of *Apis cerana japonica*, with about 400 tightly aggregated bees. (Courtesy of Masato Ono.)

The heat-killed predator, *Vespa mandarina japonica*, being gradually released by the bees. (Courtesy of Masato Ono.)

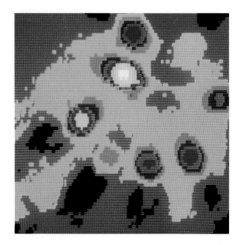

In the thermovision picture at top, white indicates the highest temperature (in this case, 40.6°C). That is the thoracic temperature of a waggle-dancing bee, which is surrounded by seven other bees. The heads of all eight bees are facing the center. The follower bees have a low abdominal temperature (black), and their thoracic temperature varies from background 34°C (light green) to about 40°C (yellow). The bee with the next highest thoracic temperature (yellow) has likely just been fed by the dancer; she will then start to shiver to warm up in preparation for her own foraging flight. (Color illustration courtesy of Sigurd Schmaranzer and Anton Stabentheiner.)

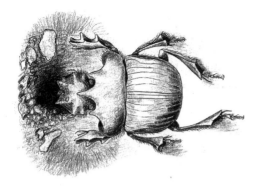

The giant earth-moving elephant dung beetle, *Heliocopris dilloni*, excavates tunnels under elephant dung piles using its blade-edged head and pushing with its powerful "caterpillar tread" legs. Weighing usually well over 20 g, it flies with a thoracic temperature of over 40°C but cools nearly to ambient temperatures when walking or burrowing in the ground.

For ball rollers, body temperature is of prime importance. The more a beetle shivers to keep warm (with its flight muscles), the higher the temperature of the leg muscles adjacent to the flight muscles in the thorax and the faster its legs can move. (It is the legs that pat the dung into a ball and that are used to roll it away when done.) Endothermy thus aids in the scramble competition for food, and it also reduces the duration of exposure to predators. Additionally, hot beetles have the edge in contest competitions over dung balls made by other beetles; in fights over dung balls, hot beetles almost invariably defeat cooler ones, often despite a wide size disadvantage.

Given all the advantages of being hot and the physiological capability of heating up by shivering, one may wonder why all beetles don't keep themselves continually hot, each trying to be hotter than the next one. As in so much of life, the answer is likely economic: we all know how to spend, but only some of us can afford it. To remain hot by shivering after landing requires a beetle to maintain a continuous and phenomenal rate of energy expenditure. It may take as much as an hour to make a large, rollable dung ball, and maintaining endothermy for this long is a cost few beetles can afford. Each beetle needs a reserve of energy for future flights and fights, in case its ball gets stolen and it doesn't soon find another dung pile. Thermal strategy is therefore of some importance, the object being to get the most dung at the least cost. In beetles that includes theft, and the timing of attempted thefts is critical.

Most individuals of both species attempt to steal partially or fully made balls when they have the thermal advantage—namely, immediately after landing at a dung pile, when they are still hot from flight. At least in *K. nigroaeneus,* the contestants respond to increased competition not by heating up by shivering to become better fighters but instead by making very *small* balls: they "cut their investment and run," so as to reduce fights. Their strategy makes sense because while making a large ball a beetle may be challenged dozens of times by a continuous succession of newly arrived beetles that, because of their higher temperature, all have the advantage in combat. Beetles already in progress of ball construction therefore have little chance of winning fights and keeping their investment; with each new arrival the chance of losing their investment increases.

The first two installments of the dung beetle story were the result of work I did in Kenya's Tsavo Park with my doctoral "father," George A. Bartholomew. It was readily published and well received. The third installment came later, from work done with my partner and former student Brent Ybarrondo, in South Africa's Kruger Park. As is usual, the study itself was much more eventful than one would guess by reading the technical report, which necessarily left out the juiciest details. In this case, however, when we submitted our report to *Physiological Zoology* (it was graciously accepted), it generated a spirited exchange of e-mail messages that perhaps revealed more than intended. Apparently one of the reviewers of the manuscript questioned our usage of the term *hominid* dung for what he thought we meant, namely baboon dung. Well, it wasn't only baboons we were referring to, because while in the field we (and the beetles) avidly made use of all human resources that we could muster. At first it was unclear what the proper inclusive term might be, but when matters were finally clarified we resolved that the PC (phylogenetically correct) term to use in our technical report was *primate,* not *hominid.*

Cold Lovers, Hot-Blooded Mates

Allusions to temperature pervade our culture, and in everyday speech *hot* and *cold* are often synonymous with good and bad. We talk of hot meals, cold showers, hot deals, cold shoulders, etc. It is not always clear if we mean temperature *per se,* but there is the vague assumption that hotter is better and cold is worse and therefore that better is always hotter. Few actual measurements have been made to test these assumptions, however. Insects—studied with scientific exactitude and with electonic thermometers—are the happy exception.

A male dragonfly may be hot when it mates, but that is only because, unlike moths, it does so in flight. As already mentioned, however, mating in moths might be called "cold sex." Even the most endothermic ones, like hawk moths and saturniid moths, settle down to a leisurely nap of some hours to mate, when for all intents and purposes they are torpid; for them as well as for most insects at rest, body temperature quickly assumes ambient temperature. It is an entirely different matter when it comes to *getting* a mate.

For large insects, endothermic heat production is a requisite for flight, and it is during flight that other activities, including foraging, oviposition, and predator escape, as opportunity and necessity dictate, may occur. Hence, to find a *direct* effect of body temperature on mating success specifically, one must examine a mating behavior that is not already tightly linked with some other temperature-dependent activity. Singing in some species is a good candidate.

Singing is one activity that serves only for mate attraction, and in katydids and cicadas only males sing and the females remain silent. Male katydids—most are green, grasshopper-like insects—make their shrill songs by rubbing their wings together to vibrate them (a mechanism analogous to that of a violin). The vigor of this singing activity (called "stridulation") is associated with and dependent on thermoregulation.

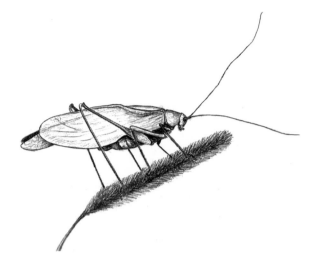

A katydid (Tettigoniidae). The males make a loud noise to attract females by rubbing their wings together, using their heated flight muscles.

Katydids of the species *Neoconocephalus robustus* warm up for their ear-shattering mating concerts by shivering, which brings flight-muscle temperatures above 30°C as needed. Other species of katydids, as well as most crickets, do not require such high temperatures for mate attraction, possibly because they sing less vigorously. The high energetic output of stridulatory singing that is achieved by *N. robustus* could be the evolutionary result of inter- and/or intra-specific competition to "out-shout" the other singers. At this time in evolution, presumably the loudest (and hottest) males attract the most females from the farthest distance.

The males of the Malaysian green bushcricket, *Hexacentrus unicolor*, offer another example of hot-blooded males that seem to have evolved by sexual selection to outshout their rivals. The males sing from dusk until well into the night, and before they sing they prepare themselves by shivering to achieve thoracic temperatures near 37°C. At thoracic temperatures of 37–38°C the males are able to achieve the extraordinarily fast wing movements of up to about 400 vibrations per second (Hz). As determined by Robert K. Josephson and associates, nerve activation and muscle contraction operate at the same frequency, which undoubtedly makes their wing muscles the fastest neurogenic muscles in the animal world. In *N. robustus,* just mentioned, a much slower rhythm of about 20 Hz is employed during shivering warm-up, and the fast contraction rhythm of about 200 Hz is then used abruptly after singing begins. In *H. unicolor,* in contrast, the frequency of activation of the muscles rises *steadily* with increasing muscle temperature during warm-up until it achieves 400 Hz, the same frequency as during singing. In both warm-up and singing, however, muscle activity is not continuous, as it is in *N. robustus,* but occurs in bursts.

It is likely that the extraordinarily high contraction rates (for a neurogenic insect) of both *H. unicolor* and *N. robustus* are possible because the muscle contracts only a very short distance during singing and requires low amounts of energy relative to the slower and much longer contractions required of the same muscles for wing beating during flight.

In general, cold muscles contract only slowly, and slow singing muscles do not produce loud songs. If male katydids are to be sufficiently noisy to attract females, their flight motors must be hot enough to achieve extremely rapid contraction cycles. For flight, however, the muscles need only contract at about 20 cycles per second. In the males some of the same thoracic muscles are used for both singing and flying, but in the silent females the very same muscles are used *only* for flight. That is, muscles in the *N. robustus* females contract only up to 20 Hz—never 200 Hz, as in males. When these muscles in the *females* are artificially stimulated in the laboratory by externally generated electrical impulses at 20 times per second, they contract and relax normally, that is, at 20 cycles per second, as expected. But if they are experimentally stimulated at 200 times per second, the successive stimulations cause the muscle to go into tetanus, because there is insufficient relaxation before the next stimulation arrives. In the *males,* on the other hand, there are discrete contractions even with stimulation at 200 Hz. In short, the same muscle that is a fast contractor (at a given temperature) in males is a slow contractor in females. This difference tightly links muscle adaptation and thermoregulation to sexual selection in these species.

The specific thermal adaptations of the same muscle can also vary over time. The dragonfly *Libellula pulchella* demonstrates both the importance of body temperature for mating success and the tradeoffs required for maximizing power output as an insect matures. The young, nonreproductive adults of this species are sit-and-wait predators that typically fly with relatively low thoracic temperature. Their flight-muscle performance as determined by James Marden in the lab at the University of Vermont does not peak at any one temperature; instead, performance is uniformly spread over a wide range of low thoracic temperatures. In contrast, sexually mature males engage in nearly continuous flight in intense territorial contests. At such times they generate a very high thoracic temperature, and they regulate that thoracic temperature precisely and within only 2.5°C from their upper lethal temperature. Thus, muscle performance of the sexually mature males is narrowly specialized relative to that of young adults that do not engage in strenuous battle. These recent results conform to predictions that were made nineteen years ago, validating our general model of insect thermoregulation.

Perhaps the best examples of the large thermoregulatory investment made specifically for reproduction may be found in those (many) large insects that dedicate their entire adult life exclusively to reproduction. In insects of this type the thermoregulatory commitment to the task is so one-sided and obvious that we hardly stop to think about it.

When I do think of it, the large saturniid moths come most readily to mind. In New England these include primarily the spectacular silk moths, the polyphemus, promethea, and the pale green luna. All have large, chunky, green caterpillars that feed on the foliage of trees in late summer, pupate inside tough cocoons of silk, then spend the winter in dormancy. When the moths emerge in the summer they do not feed. They live only for a few days, exclusively to mate and deposit their eggs on the energy reserves accumulated the year before by the caterpillar.

All of these large silk moths spend the day in torpor—they can just barely crawl if disturbed. But at the species-specific appointed time, which is only a few days of the summer, and then only at a very specific hour at night, the whole population is primed for sex. The females dispense their scent to the breeze, and males begin to shiver to warm their thoracic muscles to near 37°C. Then they fly upwind to find the "calling" females, but once they meet and begin mating they again enter torpor.

To observe this ritual more closely I recently raised a caterpillar each of a cecropia and a polyphemus moth. I left the cocoons they made outside in a wire cage so that they would be subjected to the appropriate seasonal synchronizing signals. In the first week of June 1995, shortly after the trees were freshly leaved out, both the polyphemus and the cecropia moths emerged from their cocoons. They were both females. Already, at dusk, about a half-dozen males of each species were flying all around the house, and then they landed on the wire cage with the females. Though before they had seemed like very clumsy creatures, now their flight was remarkably fast and erratic, much like that of bats. Of course, they were no longer clumsy because they were now hot-blooded, just like bats themselves. Whenever one of these moths stopped at the cage, it soon shivered, to maintain its thoracic temperature above 37°C.

I fixed a thread around the waists of my two females and tied them to a branch of an apple tree (the cecropia) and a birch (the polyphemus). The next morning each had a torpid male attached to it, in copula. Once released by the males, who presumably went on other errands, the females deposited their eggs on the branch of the appropriate food tree to which they were tied, and then they died.

In two more weeks each of the branches was alive with a horde of small caterpillars. These would begin the summer-long task of growing and accumulating the energy resources to be used for thermoregulation and flight almost a year hence, in 1996. I shall be watching for them again, to renew my wonder.

Most butterflies are much longer-lived because they feed as adults, or vice versa. Since they maintain an elevated thoracic temperature in part by basking, it is not so easy to quantify their allocation of energy to thermoregulation in part simply by weighing body lipid reserves, as is possible in silk moths. However, in a recent study Robert B. Srygley at the University of Texas—Austin has attempted to assess the cost of shivering in the reproductive behavior in two owl butterflies, *Caligo* and *Opsiphanes* (Nymphalidae), from Panama. These relatively large butterflies are unusual in that they are active only during dawn and dusk. Although these butterflies can take off and fly with a thoracic temperature of 19–20°C, they nevertheless shivered (at an ambient temperature of 24–26°C) to achieve 34–41°C in the thorax. The males aggregate at display grounds, where they wait for females and where (at least in *Caligo*) competition among the males leads to physical contact, "even to the point where males knocked each other to the ground." *Opsiphanes* merely circled one another, in apparent tests of maneuverability.

The performance of the male butterflies on the display grounds depends on their having an elevated thoracic temperature, and since the male-male contests are much more frequent than predator-prey interactions (which were not observed), it is likely that relative mating success has been the most important factor for the evolution of their thermoregulatory behavior. Srygley estimated further that the lipid reserves in the butterflies are sufficient for about forty territorial bouts in *Caligo,* and ten in *Opsiphanes.*

Every year in early spring the hover flies (Syrphidae) in my own back yard in the Maine woods employ a similar strategy. In late April, weeks before the hemlocks sprout new shoots at the tips of their dark-green twigs, and even before the beeches unfurl their pale, broad leaves, the woods of northern New England come alive with the vibrant songs of the returning wood warblers. Their sibilant melodies, so varied and diverse, are testimony to the coming of spring. But the sounds of spring include insects, too. If you listen closely you'll also hear a thin, high-pitched whine. It is a subtle sound that does not carry far. But if magnified in the mind, it can sound like a chain saw. Unlike a chainsaw buzz, though, the pitch of the sound of the fly's shivering increases gradually, eventually becoming a high-toned whine. Then

A male syrphid fly, *Syrphus* sp., waiting on a sunfleck to intercept a passing female. The alert but stationary flies keep continually heated up by a combination of basking and shivering.

. . . *Zoom* . . . ! Like an arrow off a bow, a small object shoots off a twig into a nearby shaft of sunlight, then stops abruptly in mid-air. And there it remains suspended, as if jelled into place. It remains in a shaft of sunlight that has pierced the dark hemlock grove, perhaps near a patch of snow that has remained behind green moss-covered rocks. The sun's rays play through the hovering insect's wings rowing the air like two propellers at 300 times per second. They reflect off the fly's shiny gold-copper thorax containing the powerful muscles churning out 300 contractions per second.

As we know from other studies, the hovering flies are likely all males, poised to instantly pursue any female that may come near. Among these male groups, the prize goes to those with the fastest reaction times, and the fastest flyers. That is, it goes to those who have the resources for endothermy and are able to keep hot, despite the chilly weather.

11

Thermal Arms Races

In an arms race, more is better—but not because there is any intrinsic value to the commodity (weapons) each side is racing to accumulate. To the contrary, the commodity may be of negative value, since both sides may lose economically as they add to the stockpiles of nonproductive goods. A weapon only *acquires* value because the other side has it, and the value is measured against the standard set by the other.

Arms races are common in nature. The height of forest trees is the result of an arms race initiated hundreds of millions of years ago among ground-hugging green plants, each competing against the others to reach the sunlight. The swiftness of predators and prey is the result of an arms race, as are computers and jet planes and the high body temperature of some insects. Arms races are expensive, and the lesson from biology is that they are won mainly by stint of superior tolerance or a surfeit of resources. Since thermoregulation is energetically expensive, especially for insects, one would expect to find thermal arms races an important factor in insect evolution.

The regulation of body temperature is expensive in terms of water and energy, and sometimes also in terms of time and exposure to predators. In many cases the burden of these costs, which must be borne if the insect is to be active, can be exploited to advantage by a competitor having greater resources. Over time, the escalation of resources devoted to competition may result in a spiraling arms race, leading to the ultimate exclusion of one of the competitors.

Into the Thermal Frontier

On a summer day in the Sahara Desert in Algeria, as the sun rises and begins to heat the sands that have been cold at night, an abundance of insect life is forced to retreat to cool underground refugia. Those unfortunate ones caught out in the heat become disoriented, and moving frantically they heat up even more, then they die. The sun keeps rising, and sand temperatures begin to exceed 46°C. The desert lizards, *Acanthodactylus dumerili,* continue to hunt the incapacitated prey and any ants they can find. But they now dash quickly across the sand, and when they stop and stand they alternately lift their feet to prevent burning them. Meanwhile, long-legged silver ants, *Cataglyphis bombycina,* avoid the lizards by remaining in their burrows under the sand. However, they are poised to leave, waiting for the sand temperatures to heat up even more, until it reaches about 60°C, when the temperature of the air at ant height is about 46.5°C. Temperature "testers" among them lurk at the nest entrance. At the right moment they signal by releasing chemicals from their mandibular secretions. The rest of the colony then rushes out into the field.

To the ants the heat is both their weapon and their shield. It is their weapon because it kills their prey, and they wait until the heat has left corpses for them to scavenge. It is their shield, because the lizards that would eat them retire from the field at near 46°C. It is only when even the lizards have retreated because of the heat that the ant legions rush out *en masse* from their burrows. They frantically scour the thermal battlefield only until they, too, must retire back to their underground shelters—when they experience air temperatures of 53.6°C, which is just a fraction of a degree below their thermal death point.

The hazard the *Cataglyphis* ants face is not only that of being tightly squeezed between predators and lethal temperatures. Like other ani-

A *Cataglyphus* ant, one of the most heat-tolerant insects in the world. It tolerates very high temperatures as a result of thermal arms races with predators and prey.

mals in the desert, they have to contend with a scarcity of water. A human being foolishly running at such high temperatures needs to drink several gallons of water per hour to keep hydrated. These ants, however, don't drink at all. Not having access to free water, they must make do with the meager amounts of water that is produced as a necessary by-product of metabolism. That leaves none for pouring over the body as a way to cool off. Their outer cuticle is watertight, and they don't cool by sweating. They don't pant. Instead, as they run across the desert sands, they close their spiracles, in effect "holding their breath," to reduce water loss. As a consequence, not only is water loss momentarily prevented but internal carbon dioxide concentrations build up. Using intricate instrumentation, John Lighton and Rüdiger Wehner have learned that periodically, when carbon dioxide levels reach intolerably high levels, the ants briefly open their spiracles and the dissolved CO_2 then boils out of their blood as the gas bubbles out of an uncapped soda bottle. Along with it comes a small amount of water, which they cannot avoid losing during that brief and necessary opening of the spiracles. But the silver ants are just one species of insect warriors who do battle with their enemies on the thermal battlefield.

In the southwestern deserts of the United States near Phoenix, Arizona, for example, the desert or Apache cicada, *Diceroprocta apache,* also engages in a thermal arms race against vertebrate predators, and from an evolutionary perspective it has won. These cicadas are active at the hottest time of the year, and even then they wait until the high mid-day temperatures of 44°C (in the shade) near noon to be most active. That is when they strenuously court.

When a male cicada begins to sing to attract females to it, the birds and cicada-killing wasps that feed on them have already ceased their activity because of the intense heat. These enemies of the cicada cease activity primarily because they lack sufficient amounts of a vital resource—water. Water is needed for evaporative cooling, and the cicada not only have a unique (for an insect) "sweating" response, they also have access to water that their competitors can't reach. This specific cicada is thus able to exploit the hot, predator-free environment largely because of its unique evaporative cooling mechanism (see pp. 75–77). It gets the water because it, like all cicadas, has inherited unique sucking mouthparts. Its feeding behavior allows it to tap deep, underground water stores through the phloem of desert plants such as mesquite. The cicada's competitors and predators, not having access to this water, must observe strict water economy. For that they must relinquish the field, for they cannot fight the heat without evaporative cooling.

In the deserts of southern California, the grasshopper *Trimerotropis pallidipennis* has a similar anti-predator strategy. But these grasshoppers, like the silver ants of the Sahara, endure heat rather than regulate its loss. By blending in with the background of the desert floor, they hide in order to escape bird and lizard predators. Normally grasshoppers that inhabit the ground surface stilt high above that substrate when it becomes heated to very high temperatures in sunshine. But in order to remain camouflaged it is imperative for *T. pallidipennis* to *crouch* down onto the searing hot ground. When that ground heats to near 60°C in sunshine, the duration of time that a grasshopper can remain hidden is limited by how high a body temperature it can tolerate. *T. pallidipennis* has evolved to tolerate the extraordinarily high body temperature of 50°C and can thus escape into the sanctuary of sunlight, where a predator such as a lizard or bird cannot hunt.

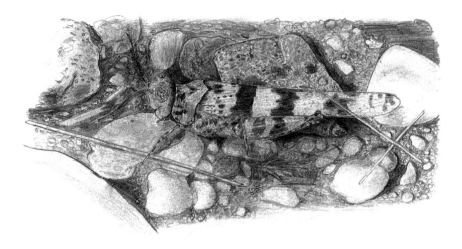

The desert-dwelling grasshopper, *Trimerotropis pallidipennis,* crouching onto the ground to escape being seen by birds and lizards. Remaining hidden in their environment requires being able to tolerate a very high body temperature.

Desert grasshoppers are not the only insects to hide in the heat. Like most other insects, ground-dwelling beetles generally try to avoid the heat. When they do not, then it is likely due to some special ecological circumstances relating to predators or food. Some tenebrionid beetles provide an exception. Most tenebrionids are cool-loving creatures, and the desert ground-dwelling tenebrionid beetles studied by Jim Kenagy and Robert D. Stevenson in Washington State fall into that pattern. They maintain thoracic temperatures averaging 20°C in the field, in part by being nocturnal during the summer, when ground temperatures in the daytime exceed 50°C, and by switching to diurnal activity in the other seasons, when ground temperatures are near 20°C in the daytime. In contrast, in southern Africa a community of other tenebrionid beetles are active on hot sand substrates at very *hot* times of the day. Inasmuch as they eat wind-blown plant debris, which would be available at all times, their "thermophilic" behavior of choosing body temperatures averaging 37°C cannot be explained on the basis of food availability. Perhaps they, too, are engaged in a thermal arms race for which they must escape nocturnal predators by venturing into the heat of the day.

Into the Cold Zone

It is probably rare that insects escape predators by seeking out *low* temperatures, but it happens. Possibly the best candidates are the Cuculiinae, a subfamily of the generally endothermic Noctuidae or owlet moths. The Cuculiinae are a northern circumpolar group of moths, and in northern New England they may fly during any month of the winter when temperatures reach 0–10°C. The remainder of the time, when they are not active, they escape the subzero temperatures by burying themselves under the leaf litter. During flight, cuculiinids have flight-motor temperatures near 30–35°C, as do other moths of their size and wing-loading. Unlike all other moths, however, the cuculiinids can begin to shiver at the extraordinarily low muscle temperature of 0°C, and they continue shivering to warm up all the way to 35°C.

The cuculiinids are unique in that their life cycle is reversed from the common pattern: other moths overwinter in egg, larval, or pupal stages, depending on the species. The winter moths lay their eggs in late winter or early spring, and by the time the first migrant birds and bats return, the larvae of most species are already partially grown. They soon mature and then burrow into the ground to pupate. They

A shivering cuculiinid winter moth (Noctuidae).

emerge as adult moths in the fall or very early spring, when many of their predators are gone. The moths in at least the northern part of their range thus escape some and perhaps many of the birds and bats that are major moth predators.

Owing to the research of the late Kenneth Roeder of Tufts University, noctuid moths are well-known for their ability to hear the echo-locating cries of bats and hence make evasive maneuvers. The ears of noctuid moths consist of air-filled compartments between thorax and abdomen that have an auditory tympanum on each side. These air-filled spaces are a pre-adaptation in winter moths that, by reducing heat leakage to the abdomen, aid them in sequestering heat in the thorax. Do they still function as ears as well?

Anne M. Surlykke and Asher E. Treat have recently examined hearing in winter moths, and their results demonstrate that the moths still have neurally responsive ears. However, those species of winter moths flying in *late* winter have significantly higher hearing thresholds at higher sound frequencies (such as those emitted by bats) than those flying earlier. This suggests that the late-winter moths would be less able to detect bats. The authors propose that hearing would be a neutral trait in the absence of bats and would not have been selected against. They therefore conclude that hearing in these moths may represent an earlier stage in the evolution of moth hearing that has been "frozen" at the level it had reached at the time when they became winter-active. That is, it apparently became easier for these moths to reduce bat predation by becoming winter-active than to stay one step ahead of the bats by improving their sonar detection.

Numerous adjustments to low-temperature survival are observed in the cuculiinid moths to permit winter activity. Physiologically the adjustments include not only the ability to shiver at very low temperatures but also the ability to mature eggs at low temperature and morphological adaptations (their thick, insulating pile and countercurrent heat exchangers) that reduce heat loss from the thorax (see p. 93). Behaviorally, the moths have also adapted by not shivering when flight is no longer necessary.

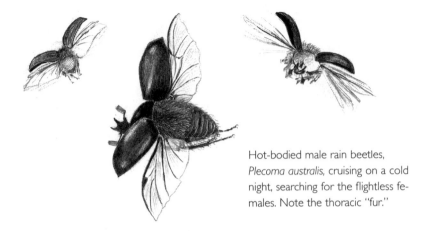

Hot-bodied male rain beetles, *Plecoma australis,* cruising on a cold night, searching for the flightless females. Note the thoracic "fur."

Among other insects that have adjusted to low temperatures are the stoneflies (Plecoptera), which are winter-active both as larvae and as adults. Plecoptera tend to inhabit cool, well-oxygenated water, and in many parts of their range the water is suitable only in the winter. The animals' life cycle has shifted to coincide with winter conditions. In summer they burrow into the stream bottom, where they remain inactive. They thus escape much predation, since fish feed very little in the winter. At this time in the evolutionary game it is difficult to say whether physiological requirements originally restricted the stoneflies to low temperatures, or whether they escaped to winter and *then* evolved suitable physiological mechanisms. Probably both processes occur simultaneously, to varying degrees in different insects. Regardless of the order, however, these stoneflies now are able to live and reproduce in environments from which they might otherwise be excluded.

Possibly an analogous case can be made for rain beetles, *Plecoma* sp., of the mountains of the western United States. Unlike most other scarab beetles that fly at moderate temperatures in the daytime, *Plecoma* beetle males fly in the winter at night, often while there is snow on the ground. Kenneth Morgan has found that, despite flying at very low air temperatures, they maintain flight-motor temperatures near 38°C, and they continue to maintain these thoracic temperatures even while digging to excavate virgin females. It is not clear why they should be active at night, when temperatures are much lower than in the daytime and hence considerably larger energy supplies are required to remain endo-thermic. An evolutionary response to escape daytime predation (birds?), if not now then in the evolutionary past, is plausible.

In most temperate regions there is not only a seasonal but also a daily cycle of temperature. Mornings are cool, then the air warms up throughout the day and cools again at night. Insect activity in general follows the temperature cycle, going up and down with temperature in the majority of small species. And this cycle provides an opportunity to some, like the hunter hornets.

The white-faced hornet queen, *Dolichovespula maculata,* comes out of the ground after a long winter's sleep in early May. She then recharges her energy supplies by sipping the sweet sap at a feeding station made on a white birch tree by the yellow-bellied sapsucker, a woodpecker that has just returned from its migration in Central or South America. Other insects as well as newly arrived ruby-throated hummingbirds also avail themselves of the tree's sweet sap, and both the woodpecker and wasps then get protein by capturing the insects that come to feed. Protein is critical at this time, when the animals need it to make eggs.

Every spring you see the sapsucker licks and also the conspicuous nests of the white-faced hornets near my cabin in the Maine woods. The hornet (wasp) nests are usually in a bush of spirea or other brush within a foot or two off the ground. At first you see a grey paper

A wasp, *Vespula* sp., in an alert but resting posture.

envelope the size of a golf ball that has a long tubular entrance at the bottom. It contains a small, horizontal paper comb with cells that are shaped much like the waxen combs of honeybees. The queen places one egg into each cell, and white, maggot-like larvae soon hatch and then grow rapidly on a diet of chewed-up insect prey. The queen wasp spends her days alternately hunting and enlarging her nest.

Throughout June, July, and August the nest grows to basketball size or larger as the queen and her continually growing number of daughters, the workers, keep adding layer upon layer of paper around the outside of the nest. As they build, they chew up the carton on the inside to make more space for larger combs, and more of them. The wasps or hornets of various species make paper by chewing the weathered fibers of wood off dead trees and limbs (and my cabin walls) with their mandibles, then moistening the mass of wood fibers with saliva.

Aside from being expert paper makers, these insects are superb hunters of winged insect prey. Their primary tactics are speed and surprise. Speed relative to potential prey is achieved by starting to hunt early in the morning, before many of their prey have warmed up. Having spent the night in the nest, white-faced hornets can get an early start because they have the advantage of warmth from the nest. Shortly after dawn groups of hunters, already at 2°C (in late summer in Maine), shiver to warm up from a nest temperature near 30°C to near 39°C, then streak out of the nest. Their high thoracic temperature gives them enough flight speed to fly over and scan the foliage as they hunt for the first insects that venture out of hiding, perhaps to bask.

The hunters catch less by chasing and pursuing than by pouncing. They do not waste time trying to decipher the difference between prey and non-prey. Instead, since any small object contrasting with the environment could be a potential insect, they pounce first and discriminate later. If what they have pounced on is a piece of inert debris, it is immediately released. But if it is an adult insect, it is crushed with the mandibles and masticated into a soft pulp that is then brought back to the nest in a round bolus that is fed to the larvae.

The hornets' own soft, defenseless larvae are, in turn, undoubtedly prized food for many birds and mammals. Furthermore, the paper carton nests that contain them are both fragile and highly visible. Yet, nests are seldom destroyed. The reason is obvious. African honeybees whose colonies have been raided for millions of years by honeybadgers and hungry hominids have become the fierce insects known as "killer bees." Similarly, a breed of very defensive hornets has evolved through natural selection. In both the hornets and honeybees, thoracic temperature must be at a certain level if a defensive response is to be made.

As anyone who has ever touched or jiggled a white-faced hornet nest knows, the inhabitants are irascible and they are easily provoked. In my experience measuring body temperatures of both white-faced hornets and the African honeybees, I'd rate the New England hornets individually far above the African "killer bees" in defensive fervor. Also, although a colony of hornets has fewer individuals to unleash upon aggressors, each individual is capable of multiple stings (whereas each honeybee can make only a single sting).

Being hit by an attacking white-faced hornet is like being hit with a fast-traveling venom-tipped projectile. That is because of its temperature. Both hornets and honeybees travel faster during an attack flight than during other flight, because they heat up first. Hornets on impact have an average thoracic temperature of 40°C, whereas those flying from the nest on a foraging trip reach only 35°C. The African honeybees are also hotter when flying to attack than when flying off to forage, which must be a pre-adaptation for a unique thermal strategy to kill predatory wasps (see p. 160). The higher thoracic temperatures during attacking flights are not a simple by-product of more vigorous flight. Rather, these higher temperatures are achieved prior to takeoff.

Winning by Quitting

The energy cost of maintaining high-speed flight while hunting for prey may of course be more affordable when solar energy (gained by basking), rather than sugar or lipid (gained by shivering), is used for "fuel." Unless the predator lives on open sunlit ground, however, basking requires the insect to remain stationary, which could compromise its hunting activity. Not surprisingly, therefore, many insect predators, including both "percher" dragonflies and robber flies, are sit-and-wait strategists: they maintain high flight-motor temperature by basking and simply wait for prey to come near them, and then they pounce. For these insects the two activities, hunting and basking, are compatible. But sunshine is not always reliable, and the sit-and-wait strategy is therefore not a good one for a social insect that has young to be fed in the nest on a continuous basis.

As one might expect, if a high body temperature is used as a weapon in the capture of prey, it may also be used as a *defense* against predators. In the evolutionary arms race between predators and prey, however, the maintenance of a high flight-motor temperature is bought with the currency of energy, and not all contestants can afford it—or, at least, one can generally afford it less than another. Those that can't afford it need other defenses. Butterflies provide an example of some of the problems encountered in thermal arms races between predators and prey.

Like most insects, butterflies are usually not engaged in any particular "task" when they fly (or when they thermoregulate to fly). Unlike the single-minded bee or wasp/hornet workers, butterflies generally feed, oviposit, chase mates, and escape predators as opportunity or necessity arises. That is, the different activities are seldom exclusive. Specific flight-motor temperatures, therefore, usually cannot be classified as a requirement for one or another function, because several or all functions must be accomplished simultaneously. What is more, a butterfly may experience the thermal consequences of predator escape only infrequently in a lifetime. Nevertheless, the occasional avoidance of death in the population can still be a potent selective pressure, even if seldom or never experienced by some individuals. Comparative studies of butterflies from the Neotropics point to the role of body temperature in predator avoidance.

It has been known since the early naturalists recorded their explorations of the tropics in the mid-1800s that some nymphaline butterflies are "fast," "erratic," and "strong" flyers, whereas the Heli-coninae and many danaine and ithaniine butterflies are "slow," "heavy," and "fluttering" or "feeble" flyers. We now know that the latter flyers are protected from bird predation either because the adults contain noxious chemicals (pyrrolizidine alkaloids) acquired from the food plants that the larvae eat or because they mimic the noxious forms. Their chemical protection allows them to fly with a body temperature some 5°C lower than that maintained by the palatable butterflies, which fly at a greater speed, presumably to make a quick escape or to discourage pursuit. An elevated body temperature and faster flight speed would not only require them to take in more food, because of their higher metabolism, it would also constrain their activity to sunny weather and to open habitat where they can bask. The unpalatable species, on the other hand, may live in shade, and because of their "low gas mileage" they are able to commute long distances to scattered food sources, despite the fact that more of their cargo space is allocated to what counts—ovaries and eggs—rather than to a larger flight motor. Indeed, experiments by James Marden of Pennsylvania State University indicate that 95 percent of the palatable species can attain greater aerial lift than the maximum values for the birds that hunt them, whereas the great majority of unpalatable butterflies get by with lower lift capacities than the birds. The in-creased flight capacity of the palatable species is due both to their morphology and thermoregulation. It appears, therefore, that the low body temperatures and the morphology that result in *slow* flight have evolved in unpalatable butterflies at least in part because they have dropped out of the thermal arms race with their predators.

Male-Male Competition

The katydids and bushcrickets that achieved flight-muscle contraction rates of up to 400 Hz (as discussed in Chapter 10) may now be seen from the evolutionary perspective. For example, it makes sense to suggest that the high volume of sound output, powered by the phenomenally rapid neurogenic muscle contraction rates of hundreds per second are the result of arms races among males to outshout each other. The resulting high body temperature may then be a secondary result of the high work effort required, and the thermoregulation followed the endothermy. Alternatively, it could be argued that the thermoregulation is the primary selective pressure: that it evolved in order to produce the unique, species-specific song frequencies that the females require for mate recognition. In order to sort out the alternatives, we must examine sound communication in other orthopterans.

Chirping at such a vigorous rate that body temperature rises sharply from the intense work effort, and that requires shivering warm-up to begin it, is likely an extravagant activity that few male troubadours can afford. Most of the orthopteran singers are strict vegetarians, and foliage is notoriously low in readily extractable energy nutrients. More data are needed to confirm this, but perhaps it is no coincidence that most of these extravagant songsters are carnivores. Perhaps they have evolved their superlative shouting ability from male-male vocal arms races that were made possible by their rich diet. That is not to say that other herbivores, such as elk, for example, are not able to do the same in their impressive bugling contests. Rather, the problem is one of size. The larger the animal, the less its relative costs of self-maintenance, but even an elephant (a *very* large herbivore) must spend most of its day eating just to remain in energy balance.

Most grass- or leaf-eating orthopteran males go economy class and stay poikilothermic, but they do reproduce, so they obviously make themselves heard well enough by the females (or else the females have developed the neural sophistication to be better detectors). That they are poikilothermic has been shown in many studies by the fact that sound frequencies during stridulation are temperature-dependent. The temperature-dependence of the song of the male tree cricket, *Oecanthus niveus,* was recognized as early as 1898, when C. A. and E. A. Bessey reported on these "thermometer crickets" in the *American Naturalist*. The Besseys followed the singing activity of individual males for three weeks at a time in their home town of Lincoln,

Nebraska, finding that on any one day the crickets all over the city chirped at the same rate. But the crickets often all changed frequency on different days, at different times of the day, and those brought indoors "began to chirp nearly twice as rapidly as the out-of-doors crickets." Finally it was decided that the crickets were not following the wand of some quirky invisible conductor but were simply calling at higher frequency when the temperature rose. They were so reliable as "thermometer crickets" that they could be used to determine temperature by the formula $T = 50 + [(N - 40)/4]$, where T is the temperature in degrees Fahrenheit and N is the chirp rate per minute. That comes out to an increase of 4 chirps per minute for every increase of 1°F air temperature.

The chirping, however, must be more than noise if it is to attract females. It must also be the right tune, which for crickets means primarily the right chirp rate, or tempo. There are hundreds of species, with many potentially inhabiting the same place, and each has its own chirp rate. It would appear at first glance that chaos could result if crickets did not regulate their body temperature in order to maintain their species-specific "tag" by which the females can identify males of the right species. Several alternative solutions to thermoregulation have been found, however. One strategy is to defer calling until the temperature is appropriate, so that the female will recognize the male's specific song "tag." Another tactic (likely the one used by bushcrickets) is called "temperature coupling," and it involves a temperature-dependent sound template in the female. That is, the female's preference changes in conjunction with changes in the chirping by the male, who is also present in the field at the same time and subjected to the same weather. These considerations suggest, but do not prove, that the metabolically costly thermoregulation is not necessary as such. Thermoregulation in this case may be a costly extravagance that, like a male peacock's tail, has been imposed through sexual selection in a spiraling arms race.

12

Heat Treatments

Many insects without sophisticated thermal weaponry engage instead in chemical warfare against potential enemies. For example, ithomiid butterflies, which are poikilothermic, slow flyers, have a noxious taste that is advertised by their brilliant colors. Given their all-too-obvious chemical defenses, they require few additional anti-predator strategies and are thus able to survive without the flight capabilities endowed by endothermy. Not all insects, however, rely solely on well-advertised bad taste to be left alone. Ladybird beetles (Coccinellidae) and blister beetles (Meloidae), for example, are relatively conspicuous animals that feign death when disturbed. These slow beetles, which seldom bask and never shiver, also exude bright yellow droplets of poisonous blood from their leg joints. Their arsenal of behavioral and physiological responses deters all but the most hungry of predators.

Perhaps the ultimate defense is mounted by a nondescript little poikilothermic beetle known as the "bombardier beetle" (Carabidae). Bombardiers, when disturbed, shoot a well-aimed spray of noxious quinones at their antagonists. This capacity could arguably classify them not only as chemical but also as thermal warriors, since their caustic spray is made all the more irritating because it literally explodes from the rectum after being heated to 100°C! Cornell biologist Thomas

Eisner photographed and measured the beetle's puff of heated poisonous gases and determined that what appears to us as one puff is actually a rapid-fire volley of many puffs.

The key to the mechanism is control over the mixture of two chemicals stored in two different chambers. The two chemicals are inert by themselves but highly reactive when combined. An inner chamber contains hydroquinones, precursors of the final product. When attacked, the beetle opens a valve that releases some of the hydroquinones into an outer compartment, the reaction chamber. Here, enzymes decompose hydrogen peroxide to water and oxygen, which oxidizes the explosive reaction. There is an audible "pop" as the hot reactants then exit from the rear. The tip of the abdomen has a nozzle that can be, and is, aimed for maximum effect.

Insects must engage in life-and death struggles on the thermal battlefield not only against predators and prey but also against disease organisms. Some insects, like other animals, combat disease organisms with elevated body temperatures. Since during a "fever" body temperature is regulated as before, but at a higher temperature set-point, we may feel cold even if our body temperature is above the normal set-point, and we attempt to raise our body temperature to the new, higher one, by shivering. Similarly, nonendothermic animals with a low body temperature, including some lizards, try to warm up by basking when they are infected with pathogenic bacteria.

High temperature fights infection in numerous organisms. Peach trees suffer viral infections that are cured under the high temperatures normally encountered in summer. The caterpillars of *Bombyx mori,* the commercial silk moths, and those of many other moths and butterflies succumb to a number of viruses, but they are "cured" if reared at moderately high temperatures—near 35–38°C. The western United States rangeland grasshoppers, *Camnula pellucidae,* normally maintain body temperature at 38–40°C in the field. These temperatures speed up their developmental rate and also protect them from the fungal parasite *Entomophaga grylli.*

There are very few examples, however, of an insect host modifying its behavior so that it runs a "fever." Nevertheless, in a handful of studies, infected insects maintained higher body temperatures than uninfected ones. These include the Madagascar cockroach, *Gromphadorhina portentosa*, injected with heat-killed *E. coli*; crickets, *Gryllus bimaculatus*, injected with their natural rickettsia parasite, *Rickettsiella grylli*; the American migrating grasshopper, *Melanoplus sanguinipes*, infected with protozoan spores of *Nosema acridophagus*; and the desert tenebrionid beetle, *Onymacris plana*, injected with purified lipopolysaccharide isolated from *E. coli* bacteria.

In those species where heat treatment aids survival, the heat usually does not kill the pathogen outright, in the way that pasteurization sterilizes milk. Instead the host somehow keeps the pathogens from replicating, and suppression lasts only as long as the temperature remains elevated.

Killing internal pathogens by fever is not easily demonstrated, and numerous problems remain to be solved before we can begin to understand the possible role of fever in insects. It can be expected, however, that fever would not be a general phenomenon in insects because, although many virus infections can be reduced with heat treatment in the laboratory, an insect that exposes itself to sunlight in order to run a "fever" runs a high risk of being eaten by a bird. The insect could, alternatively, run a fever by shivering instead, but then only the thorax would be heated and the cooler abdomen would remain a haven for the pathogens. Furthermore, some of the temperatures that lower the incidence of viral infections simultaneously *raise* the risk of bacterial infections.

Although increased body temperature is the most commonly known strategy to combat parasitic infections, Swiss researchers C. B. Müller and Paul Schmid-Hempel report an insect host altering its behavior to give it a low-temperature defensive strategy instead. Bumblebees are commonly parasitized with the larvae of conopid flies (Diptera; Conopidae). A conopid fly's larva burrows into the bee and lodges in the abdomen, where it grows to a large, plump maggot that eventually fills out the abdominal cavity and kills the host. A fly parasitoid usually takes 10–12 days to kill its host, when the larva pupates. The growth rate of the fly larva depends on its temperature, which is the same as that of the bee's abdomen, hence a larva that lives in a warm bee will grow faster and kill it much sooner than one lodged in a cool abdomen. In what may be a defense strategy, bumblebee workers that contain a conopid maggot maintain a *lower* body temperature. They keep cool by reducing endothermy, as well as by choosing lower temperatures in an experimental temperature gradient in the laboratory and by not returning to the nest at night, where they would ordinarily be warmed passively. Their behavior is not without cost, however, since it prevents these bees from heating the brood and perhaps from starting to forage early the next morning at low ambient temperatures.

A difference in temperature sensitivity between an animal and its enemies has also been used to advantage by the Japanese honeybees, *Apis cerana japonica,* in what amounts to social warfare. Honeybees typically defend themselves against enemies by pursuing and then stinging them. The African race of the honeybee, *A. mellifera,* is famous for this tactic. Perhaps not surprisingly, I learned in Africa that at hive entrances attacking bees stimulate others to attack also, and that the attackers are hotter and fly faster than those bees leaving the hive only to forage. The honeybee *A. cerana* has evolved still further in its defensive response to its enemies, the large, heavily armored predatory wasp, *Vespa simillima.* The wasp hunts for bees at hive entrances, but if it ventures too close to the entrance of a hive, then the hive defenders may rush out and attack it *en masse,* as do African honeybees. As many as 180–300 bees will form a ball around the wasp, in the center of which temperatures soar to near 46°C and are maintained at that level for 20 minutes. Bees in a ball will naturally increase their temperature since surface area exposed to cooler air is reduced, if they continue to produce heat. These do. The heat treatment is lethal to the wasps, but the bees tolerate 48–50°C. The wasp is released when dead (these bees' stingers do not penetrate it).

The European honeybees, *A. mellifera,* that have been imported to Japan also respond by forming a ball around the attacking wasp, but average temperatures inside their balls are only 43°C. In these balls the hornets are killed by venom, not heat. Each hornet usually has two or three stings embedded in it, each one representing the sacrifice of a bee (honeybees die after stinging).

Since this balling behavior was first reported in 1987, Masato Ono and collaborators in the Faculty of Entomology at Tamagawa University have deciphered ever more sophistication in the attack strategies of the wasps and the thermal counterstrategies of the native Japanese bees. They determined that the giant hornets, *Vespa mandarinia,* a close relative of *V. simillima* wasps, also attack honeybee colonies. Like *V. simillima,* giant hornets can be dispatched one by one in balls of shivering bees. As in all warfare, however, counterstrategies are bound to evolve to unleash an arms race.

During a typical giant hornet attack, a lone hornet forager first captures bees at the periphery of the bees' nest. It macerates the bees' bodies and feeds them to the larvae at its own nest. After several successful foraging trips to the beehive, the hornet deposits a marking pheromone at the hive entrance from the van der Vecht gland at the tip of the abdomen. This pheromone attracts other hornets from the home nest, and then the slaughter phase of the hornet attack begins: 30,000 bees can be killed in three hours by a group of 30–40 hornets. Subsequently the hornets may occupy the hive itself, and then they carry off the bees' larvae and pupae to feed to their own young.

This is what happens when hornets attack colonies of the introduced European honeybee, *Apis mellifera,* but the Japanese honeybee, *A. cerana,* has evolved an effective counterstrategy to the hornets' mass invasion. The Japanese bees guarding the nest entrance return *into* the nest after detecting the hornets' marking pheromone. By doing so they lure the first hornet attackers or scouts into the nest, where they are killed. At the same time, the *A. cerana* nest inhabitants are alerted, and over 1,000 workers leave the combs in the nest interior and poise themselves vigilantly just inside the entrance. Those unfortunate foraging hornets that are recruited by the pheromone and then try to enter the hive are met, balled, and killed by heat. The interior of the balls quickly rises to 47°C, killing the hornet but not the bees, whose upper lethal temperature is 48–50°C. The heat of the ball may help the honeybees to volatize and broadcast their own recruitment pheromone, isoamyl acetate (which smells like banana). The honeybee pheromone recruits ever more bees from the nest interior to defend the nest.

If the bees fail in killing the first hornet attackers, then hornet nestmates instead are able to reinforce their attack and by sheer numbers can overpower the bee colony. The chemical recruitment of the giant hornets to counter the strategies of the thermal warrior bees is employed against other insects as well. The hornets also attack the wasps *V. simillima* and *Polistes rothneyi*. The *Vespa* congener mounts a mass defense to a mass attack. The *Polistes* wasps escape, never to return to the colony; instead, they construct new nests and begin the nest cycle anew.

Fighting Heat Stress

Being heated to near 50°C is a serious temperature stress to most insects, especially if they are denied all behavioral and physiological options for escaping it, as a wasp or hornet trapped inside a ball of bees is. A fruit fly larva stuck in a banana at 30°C may also experience heat stress, but it is able to use one response—a biochemical one—as a last resort.

A major problem that occurs when tissues heat to temperatures beyond those at which they normally function is that the protein molecules found in cells begin to unfold or come apart. Proteins are chains of amino acids that have distinctive three-dimensional shapes. Arrangement is crucial for these molecules: no structure, no function. Furthermore, with changes in shape reactive groups of the molecule may become exposed and form inappropriate alliances with other proteins; it is then difficult if not impossible to disentangle the newly formed molecules and restore the original proteins to their normal functioning. It is here that a group of defensive proteins, called "molecular chaperones," intervene. They guard the stressed proteins and help prevent them from unfolding, possibly by having an affinity for (a tendency to bind with) those proteins that are in the process of unfolding. The chaperones are thought to hold proteins in the embrace of a barrel-like structure that prevents their charges from unraveling and behaving inappropriately.

The chaperones are also more commonly called "heat shock" or stress proteins because they were first found to be induced (in *Drosophila* larvae) in response to heat shock and other stress. (Their functional role as chaperones was only discovered much later.) The study of these proteins has undergone an explosive growth since the characteristic pattern of puffing on chromosomes (indicative of protein synthesis associated with specific genes or groups of genes) was first discovered in response to heat shock in 1962. There are now some 1,000 papers annually on the topic. All organisms face stress and a great many of them at some time face unavoidable overheating. Heat-shock proteins are found in organisms from plants to humans, and, as they prevent heat-sensitive proteins from unraveling, they typically improve an individual's tolerance to high temperature.

The role of heat-shock proteins (HSP) in the thermal tolerance of fruit fly *(Drosophila)* larvae has recently been examined using molecular techniques. In *Drosophila* there are at least ten copies of the *hsp70* gene (as opposed to only one copy of the *hsp104* gene in yeast). Genetically engineered fruit flies were created in an experiment by inserting twelve extra copies of the *hsp70* gene in *D. melanogaster*. An "excision" strain has also been created that lacks the extra copies of the *hsp* gene. As one would expect if the *hsp70* aids in thermal tolerance, fruit flies from the extra-copy strain of *Drosophila* show more rapid production of heat-shock proteins in response to heat stress than do flies from the excision strain, and the high-HSP strains have much greater survival rates in response to high-temperature stress. Apparently the insects do not constantly maintain very high amounts of protective heat-shock protein because large amounts are deleterious to cell growth and division in the absence of stress. It is therefore adaptive to get rid of the chaperones when they are not needed, and to induce their production only in response to stress.

13

Heating and Cooling the Nest

There are a number of ways of maintaining a comfortable temperature in a home. You can build partially underground or otherwise carefully pick the house site. You can exploit the energy of the sun with solar heating. Or you can add insulation or install a furnace. Social insects use all these means to varying degrees, and they also harness the heat of their own metabolism to control the temperature of their surroundings. But making it all work has required numerous innovations in their behavior. Over millions of years, social insects have found ingenious ways of heating their homes, and we could learn from their examples.

Nest Site Choice

Finding the proper nest site is perhaps the most important aspect of colony thermoregulation for those species that have only a limited capacity for regulating colony temperature. Ants are a primary example. Colony success of ants in north temperate regions is often determined by the thermal environment of the site the queen selects for her colony. In northern Europe, for example, colonies of the common red wood ant, *Formica rufa,* are generally not successful unless they are founded at the edge of a clearing where they receive direct solar heating. Those colonies that are started in shade grow very slowly and produce no reproductives (new queens and males) before dying out. At high elevations in the Alps, on the other hand, ants of various species nest under small, flat rocks that heat up in direct sunshine, whereas they avoid the nearby vegetation-covered soil, which remains cool.

In deserts the problem is primarily high rather than low temperature, as well as huge fluctuations of air temperature. Soil temperatures are fairly stable, although they vary vertically. Appropriate nest location is less dependent on the location of the site than on its depth beneath the ground surface. In the Sahara Desert, nest chambers may be several meters below the surface, where temperatures stay cool. Seasonally, the ants may change nest sites or choose appropriate temperatures by occupying different depths of the same nest. Many solitary and semi-social ground-nesting bees face a situation similar to that of the ants: in the northern hemisphere they, too, site nests in accordance with the thermal environment of the ground.

Bumblebees rely on their own heat production for maintaining a high nest temperature (near 30°C). But in order to do so at low air temperatures, they require insulation. Thus, queens that emerge in the spring spend considerable time and energy searching for a nest site with access to material such as fur, feathers, or dry, shredded vegetation. In early spring, soon after the snow melts,the queens skim close to the ground and land at frequent intervals, examining holes and burrows for abandoned mouse nests. In the High Arctic, *Bombus polaris* queens use lemming nests and also the freshly made feather-lined nests of snow buntings, *Plectophenax nivalis*. Apparently they will even evict incubating birds from their nests.

The different species of tropical honeybees of the genus *Apis* build uninsulated comb nests hanging in the open. Local nest site selection is of relatively minor thermal importance to them, and some may migrate seasonally between sites as prevailing ambient temperatures change.

For the European race of the honeybee, *A. mellifera,* nest sites are permanent and nest site selection is the key element for surviving in the northern hemisphere. When a cavity is not available, these bees ultimately resort to building their combs in the open, like the tropical *Apis* species do, but these nests invariably perish in the winter. Proper nest site selection is achieved by a highly integrated social response.

When a swarm (over ten thousand workers plus the old queen) leaves the colony, it typically hangs in the shape of a beard, and scouts leave from it to scour the countryside for suitable nest sites. The scouts inspect numerous cavities to gauge their quality. Several criteria are involved in nest site selection, including cavity size and entrance hole (preferably small and south-facing). Several scouts may each find different potential nest sites, and each scout then indicates the apparent suitability of its best find by the vigor of its direction-indicating dance. Scouts also follow each other's dances, calibrating the quality of their own finds versus those of others; they will be "converted" if they encounter a better one than they had themselves found. Eventually all the scouts reach a consensus, and the swarm then departs *en masse* to the best nest site that any of its scouts had managed to locate. "Streaker" bees who know the nest site location by already having flown individually to the site indicated lead the way. The queen follows.

The European honeybee is of tropical origin, and after many thousands of years in northern climates it still needs to maintain nest climate of near 32°C throughout the whole year. In order to maintain that tropical climate in the winter, honeybees consume large quantities of honey to produce heat. Honey stores are thus a key to nest climate control, as is a good supply of oil or wood for us. If a colony is to lay up sufficiently large stores before winter it needs an early spring start.

Colonies divide by fission, and each offspring colony starting without stored reserves needs close to the entire summer to gather sufficient fuel to survive winter. Thus, late swarms are as good as dead swarms. (It is, of course, the bees' adaptation for storing honey that humans have exploited, primarily by stimulating them to produce more than they need and then taking the extra.)

In order to issue early swarms, the parent honeybee colony must start rearing young while it is still winter, using the stored food laid up the previous summer. The brood is very sensitive to temperature, and developmental abnormalities, such as wrinkled wings, result if the brood temperature drops below 30°C. In general, brood temperature is maintained near 32°C, even as outside temperatures drop to as much as 80°C lower than that—as may happen in late winter or early spring, during the crucial early brood-rearing phase that prepares the colony for the early spring honeyflow. Obviously nest site selection is extremely important to permit such impressive thermoregulation necessary for the seasonal colony cycle.

Nest Construction

Social insects are well known for their sometimes elaborate nests. Temperature control in these structures is a major factor in the survival of some species, although it is of minor importance to others. As in all aspects of insect ecology, nest structure is related to environment and phylogenetic heritage.

Ants do not generally make structures that function in nest thermoregulation. But the large mounds of the northern European wood ant, *Formica* sp., may be an exception. The big red wood ants construct their mounds by accumulating conifer needles, short sticks, and other organic debris. Mounds may be 1–2 m tall and are placed so as to be exposed to sunshine for at least a portion of the day. They function as collectors of solar heat, and the greater their mass, the more heat they can store and the slower they cool through the night. The heat collected during the day lasts most of the night. In addition, bacterial decay within the mound generates warmth, as it does in a compost heap.

Precise regulation of temperature in any one spot in an ant nest is not necessary, because the eggs, larvae, and pupae are loosely distributed into piles within the nest, and they are carried by the workers to wherever in the nest temperatures are appropriate. Thus, in the night the ants bring their brood deep into the interior of the mound, while in the morning, when the sun shines on the mound, they bring it close to the now warmer surface layers. As a result of this behavior, brood temperature is regulated within 1–2°C.

Different bee species employ varying degrees of nest construction for thermoregulation. Bumblebees "card" or fluff out the already existing nest material, such as used mouse nests that they take over and use, to make it more effective insulation. They may also pull additional insulating material into the nest from the nest surroundings and then construct a wax covering or roof over the brood. Honeybees similarly plug extra holes or reduce the size of the hive entrance hole with plant gums (propolis). Some stingless bees (*Trigona*) build enclosed nests of carton onto open branches. In one species the nest is equipped with numerous ventilation holes that are opened or closed depending on temperature.

In the northern hemisphere vespine wasps, commonly known as hornets or wasps and yellowjackets, build hanging horizontal combs in which broods are suspended in cells. These combs are surrounded by several layers of paper, or "envelopes." Since warmed air produced by the nest inhabitants rises, the nests' ventral entrance hole is ideally situated to reduce heat loss. Many vespine wasps also gain additional insulation by building their nests in the ground or in other sheltered locations, such as tree cavities or in or on our own dwellings.

Although most species of termites have no known climate control and no conspicuous nest architecture, nowhere is nest architecture and climate control raised to such sophistication as in some of the higher termites. Like the ants of desert regions, many termites build entirely subterranean nests. But termites have also colonized areas where flooding occurs regularly, and here the species have evolved impressive structured domiciles that reach far above the ground. These structures are then subject to the full brunt of the tropical sun, and they are built in ways that enhance climate control.

Unlike the mounds of ants that consist largely of loose debris, those of termites are hard, cemented structures built by the accretion of tiny packets of moist soil, such as clay, mixed together with feces to produce mortar-like walls. They are built by individuals carrying, and depositing, one packet at a time.

The resulting shapes of termitaria vary enormously, and they are custom-designed for specific environmental contingencies. For example, *Cubitermes* termites from rainy areas of western Africa build mounds with mushroom-like caps that function as rain parasols. Conversely, *Macrotermes subhyalinus* from arid East Africa construct open chimneys up to several meters tall for the release of heat produced internally both by the collective metabolism of its several million inhabitants and that of their fungus gardens. Air passages through the chimney-like structures are connected with openings leading into the nest near the ground, so that air flow over the tops of the chimneys can suck air through the system and remove heat.

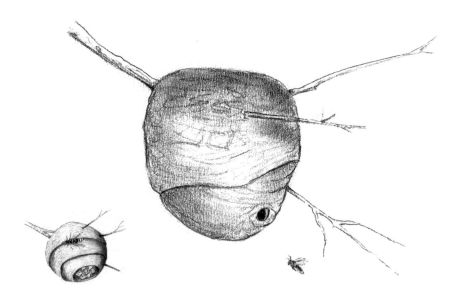

A wasp, *Vespula* sp., nest is produced by adding successive insulating layers of paper around a comb. As the colony grows the nest is enlarged *(top)* and a small entrance hole remains at the bottom of the nest.

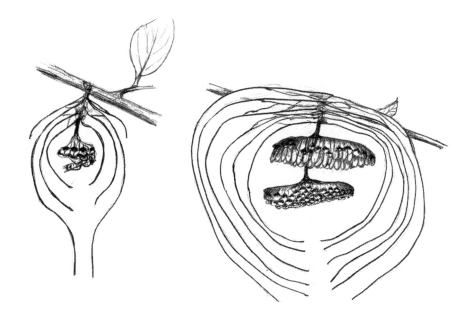

Cross-section of a vespine nest. The queen nests *(left)* of the white-faced hornet, *Dolichovespula maculata,* often have a long ventral entrance tunnel, which may reduce heat loss. Mature nests *(right)* lack the entrance tunnel but have many more paper envelopes and more combs.

The ultimate of termite engineering may be reached by the African termite, *Macrotermes bellicosus*. The metabolic heat from the millions of inhabitants and their fungus gardens also rises in the nests of these termites. But rather than exiting from chimneys at the top, the air is redistributed into numerous narrow tubes that run in flanges alongside the mound exterior. Gas and heat exchange takes place in these flanges as the air recirculates. As worked out by Martin Lüscher, in this "thermosyphon" system the air rises in a relatively closed system, therefore creating a negative pressure behind it that sucks air from the descending external air tubes back into the base of the territory. The syphon that drives the air circulation is caused by the thermal disequilibrium. In this system potential heating on cold mornings could

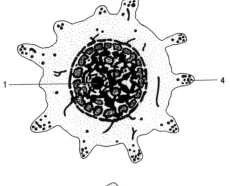

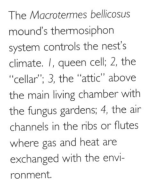

The *Macrotermes bellicosus* mound's thermosiphon system controls the nest's climate. *1*, queen cell; *2*, the "cellar"; *3*, the "attic" above the main living chamber with the fungus gardens; *4*, the air channels in the ribs or flutes where gas and heat are exchanged with the environment.

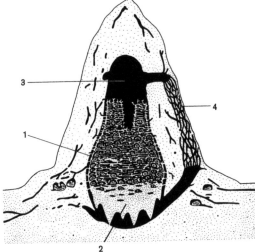

also result in the air flanges acting as solar panels. Few data are available, but apparently some automatic temperature stabilization is due to the mound morphology. Additional control is added by the termites themselves, however, because when all of the inhabitants are killed (by poisoning) then fluctuations in the internal mound temperature increase. The mechanisms of temperature control in this system are still not well understood.

Nest temperature stabilization is relatively automatic in the Australian "compass termite," *Amitermes meridionalis*. This species builds steep (2 m or taller), wedge-shaped mounds aligned along the north-south axis. In this orientation the mounds receive maximal exposure to the sun in the early morning and late afternoon, as the sun strikes in turn their broad, flat vertical walls first on the east and then the west. But at midday the sun's rays coming from overhead and the north (in Australia) glance off the steep sides at the pointed north end of the wedge, so solar heat input is then reduced. Mounds that are sawed off at the base and re-oriented east-west show much larger internal temperature fluctuations, because they are heated very little in the morning and evening but are heated a great deal during the midday, when they present a relatively large lateral surface to the sun.

The mechanisms whereby the individual actions of millions of independently acting termites of any one colony build such very differently shaped mounds are not known. It is known, however, that compass termites respond to magnetic cues, and that various termites respond to local noxious stimuli, such as light or high temperature, by depositing their packets of construction material, as if trying to create a barrier to these stimuli. Clearly the individual workers cannot have a clue as to either their existing or the final structures. They can only

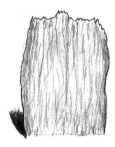

Two views of the "compass mounds" of the *Amitermes* termite of the Australian desert. The design facilitates the maintenance of a high and stable internal temperature.

respond to local stimuli. Do the different termites create such distinctive mounds because they respond to the same stimuli according to different behavioral patterns, and/or are they responding to signals and to different environmental stimuli according to a single repertoire of behaviors?

Nest Heating

Depending on the number, size, and activity of the nest inhabitants, the heat generated as a by-product of the normal activity of the colony can either be too much or too little for optimal nest temperature. In those species, like wasps and bumblebees, whose colonies are started by single queens and then build up to hundreds or thousands of individuals, both insufficient and excess heat can be problems.

Ant workers do not have thoracic muscles for shivering, so heat production specifically for nest temperature regulation is not expected. In very large mounds of the wood ants *Formica rufa* and *F. polyctena,* however, nest heating by only the "passive" or resting metabolism of the ants can occur when, in the early spring, all of the millions of individuals crowd together deep in the interior of the mound. Heat is then little dissipated by convection, and in these clustered individuals body temperature rises, generating even greater rates of resting heat production in a self-reinforcing spiral. Similarly, Neotropical army ants *(Eciton)* in the center of a large raiding cluster will experience an increase of several degrees in body temperature. So far there is no evidence for any ants actively producing heat *for* temperature regulation, but perhaps in some species clustering is a mechanism for manipulating resting metabolic rate (by reducing surface area for convective heat loss). Numerous other adjustments to existence in a northern climate are found, and much of the behavior and physiology of thermoregulation is necessary only because an insect's life cycle does not allow it to *avoid* thermal problems.

Queen bumblebees must initiate their colonies in temperate and Arctic regions soon after snow melt, while the ground is often still

A bumblebee queen, *Bombus vosnesenskii,* incubating her brood clump (with eggs, larvae, and pupae) while facing her honeypot. She shivers with her flight muscles in the thorax and transfers heat via the blood to the abdomen, which provides the contact with the brood.

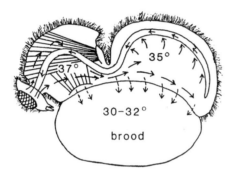

frozen and there are still frosts at night. The burden of nest temperature regulation in the early stages of colony founding falls largely on massive heat production by shivering. After making an initial wax-covered brood clump containing a pollen-honey mixture and her first batch of eggs, the queen incubates this brood clump whenever she has time left over from foraging, much as a bird incubates its eggs. She continues to incubate her brood clump while the eggs hatch, the larvae grow, and the pupae develop. As the colony builds up, there are more and more bees to help incubate subsequent batches of brood, and the heat generated as a by-product of other activities in the nest soon suffices and increasingly less time and energy needs to be expended for thermoregulation.

Although incubating bumblebees produce heat by shivering with their flight muscles, as in pre-flight warm-up, they do not distribute heat randomly in the nest. Instead they incubate the brood specifically, by applying their ventrally bare abdomen directly and snugly onto the brood. They then cling to the brood closely and as they shiver one sees rapid vibrations of the abdomen. These abdominal vibrations are the breathing movements that ventilate the working thoracic muscles, but at the same time they serve to pump hot blood out of the thorax into the abdomen so that the brood under it can be heated (see p. 96).

Honeybees, because there are tens of thousands of individuals in a colony, do not need to incubate specific brood cells locally. Nevertheless, they also do not heat the whole nest indiscriminately. Instead, they crowd into that area of the nest where the brood combs are located, and they thereby selectively heat that general area. Most honey stores are placed into the combs at the periphery of the nest, where they also serve as insulation, while the brood is kept in the nest center. Nest temperature in the vicinity of the brood is maintained precisely, within about 1°C, of 32°C, even when temperatures outside the hive in late winter may dip to −40°C (or rise up to 60°C in summer).

Honeybees must also keep warm in the absence of brood, because unlike bumblebees they do not individually tolerate temperatures near the freezing point of water. Bees in a broodless winter cluster or in a swarm cluster crowd together as a first response to a drop in ambient temperature. This is little more than an "every bee for itself" response; individual bees crowd closer to other bees, or closer to the center of a cluster, to keep themselves warm. Every bee appears to look out for her own selfish interest of keeping warm either by huddling or shivering, but as she does so she helps the whole hive keep warm as well. It reminds one of the words of Bernard Mandeville in *The Fable of Bees: or Private Vices, Publick Benefits* (1714): "Thus every Part was full of Vice, yet the whole Mass a Paradise."

There is, however, a limit to the percent of bees that can fit into a cluster: some bees are inevitably left on the periphery. The bees on

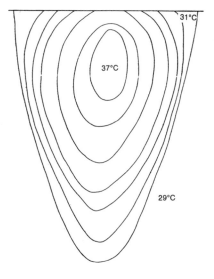

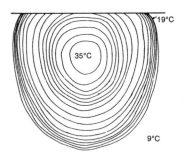

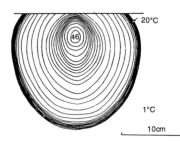

A honeybee swarm is a group of bees that has left the nest and is at an interim location before finding a new suitable nest site, such as a hollow tree. The temperatures of the core and mantle of this mass of 10,000 to 30,000 bees are regulated at quite different but relatively precise temperatures through a relatively simple set of responses of individuals that are expressed because of the social context. In the diagrams at left the lines are *isotherms* outlining areas of the same temperature; the area with the highest temperature is the core, and temperatures decrease in relatively even increments with greater distance from the core. The isotherms shown here model a captive swarm and its contraction as the ambient temperature was lowered from 29°C to 1°C.

At high air temperatures bees at the mantle of the swarm *(top)* face out and try to avoid contact with each other. At low ambient temperatures *(bottom)*, they crowd into each other seeking warmth, hence blocking all exits for heat from the swarm.

the swarm periphery or "shell" must tolerate lower body temperatures, whereas those in the center of the cluster may be subject to overheating as the shell bees crowd ever closer together. As the outer bees crowd in, the resting metabolism of the core bees necessarily increases because they become hotter. Meanwhile, those bees on the periphery may shiver to keep their own body temperature at least 14°C, so that they can remain coordinated enough to hang onto the cluster. (If a bee's body temperature dropped below 10°C, she would lose coordination and drop from the cluster.) Thus, in a broodless cluster at moderately low temperatures the inner bees may be maintaining the maximum tolerable temperature while those on the outside of the cluster are at the minimally tolerable temperature. In other words, a steep temperature gradient exists in the cluster, and since the temperature gradient is low on the outside, considerable energy savings result. Equilibrium is maintained and the colony's core temperature regulated relatively precisely as the overheated internal bees crawl out and the chilled inside bees crawl into the center.

Although the primary or direct advantage of colony thermoregulation in most insects is likely the incubation of brood and perhaps also the control of pathogens, the social group can also provide numerous other advantages. The warmth of the crowd may be a sanctuary from the external environment. It may allow foragers to leave early on cool mornings, and daily fluctuations of colony temperature may serve as a "clock" that synchronizes the circadian rhythm with external time—which is very important for bees that must keep to the flower's daily schedule of nectar production. Ultimately, however, thermoregulation of the domiciles has permitted some species (such as honeybees, and humans too) to invade cold as well as hot environments.

Nest Cooling

People living in hot deserts have learned to employ some of the same architectural principles for controlling temperature in their domiciles as insects do. Undoubtedly, however, we will do even more when energy for temperature control becomes more expensive.

One principle of nest construction is use of the ground as a means of absorbing heat. The subterranean dwellings of desert ants and termites are constructed so that the surrounding ground acts as a heat sink. A drawback for humans trying to exploit this principle is principally our sometimes limited capacity to burrow and excavate; another problem is posed by the lack of natural lighting and ventilation. These drawbacks are eased somewhat by building into the side of a hill.

Another means of absorbing heat would reverse the principle of mound construction—in other words, to design a tower that directs heat away from, rather than toward, the living quarters. An example would be the main building of the Jacob Blaustein Institute for Desert Research in Sede Boker in the Negev Desert of Israel. In a recent heat wave, when temperatures reached the upper forties (over 120°F), the building remained at a comfortable temperature without any assistance from air conditioning. A massive stone column that is embedded in the center of the building absorbs heat from surrounding rooms, thus acting as a heat sink and dampening large temperature fluctuations. (It would, of course, also release heat and maintain elevated room temperatures when ambient temperature falls.)

External and internal heat sinks never provide "perfect" temperature control in nests, in the sense of constantly maintaining one optimum temperature. Environmental temperatures fluctuate, and crowding from an increasing nest population often adds too much additional heat. Thus, most social insects with very high population densities (millions of insects in the confines of a small space) need ways to dissipate extra heat. Dispersion is the first response; fanning is used to dissipate heat by those species that have wings; and, finally, for some species under extreme heating conditions, evaporative cooling may also come into play.

For example, tens of thousands of relatively large and metabolically very active honeybees may share a nest. Passive cooling of the colony—namely, by individuals avoiding the most crowded confines of the nest—can go only so far. The passive convection that results from air passages created by honeybees avoiding the highest temperatures is augmented by more active processes.

The extent of cooling that honeybees can achieve was demonstrated by Martin Lindauer from the University of Würzburg. Lindauer placed a bee hive in full sunlight on a lava field in Salerno, southern Italy, and determined that the bees could maintain close to their normal hive core temperature despite an outside temperature of 60°C. Only evaporative cooling can bring the nest (or body) temperature below ambient temperature. Not much is known about the individual responses that result in regulated evaporative heat loss, but the general pattern involves the socially regulated intake of water, which Lindauer studied, in conjunction with fanning.

Bees normally deposit nectar into the combs above the brood nest. As water evaporates from this nectar to concentrate it into honey, the air becomes humid. The bees then regulate the relative humidity of the hive by fanning, which drives the water-saturated air out of the hive and brings in drier, unsaturated air. When the air inside the nest is less humid, more water inside the nest will evaporate. Thus, the more water or nectar is deposited in the hive at high temperatures, the more the bees must fan and the more the evaporative cooling that occurs.

Honeybees align themselves into existing air streams while fanning, thereby moving air *through* an overheated hive.

Fanning by honeybees is not merely a stirring of the air in the hive, such as the ceiling fans in our dwellings accomplish. Fanning bees preferentially align themselves along already existing airflows and so unidirectional air flow is created, especially where it is least impeded and hence the most appropriate. Apparently gangs of fanners first stir the air in the broodnest, and then the air moves along lines of least resistance, where it is moved farther by the chain gang. Eventually the air exits, usually at the hive entrance. At least it is there that we find the pleasant honey scent during a good honey flow, and presumably during hive overheating.

Fanners do not necessarily themselves perceive the hive temperature, since at high ambient temperatures many returning bees stop to fan at the hive entrance before entering. These bees tend to be primarily the pollen gatherers, who have higher thoracic temperatures than the nectar gatherers (because they have less nectar for evaporative cooling; see p. 73). Possibly, therefore, a high thoracic temperature itself, if coupled with appropriate hive stimuli, stimulates fanning behavior. If so, it is tempting to speculate that this behavior evolved from active convective cooling of the individuals, with the response becoming modified for the social context. Another example of an individual behavior modified for social benefit is "tongue-lashing," where regurgitated nectar evaporates from a bee's tongue. The tongue-lashing behavior first reduces the individual bee's own body temperature (see Chapter 6), then necessarily also that of its environment, the brood nest with young.

Many of the bees' responses are shaped by social needs, and in that sense the colony is like a machine, or "superorganism," with different parts that contribute to the functioning of the whole. Although this analogy acknowledges the obvious, it is not generally useful in specifying how coordination is achieved. It does not reveal how the individuals react, or why they react, in the way that we know how the various organ systems are controlled to produce an overall effect in an individual organism. Ultimately, the functioning of the whole can only be understood by dissecting it and learning how, when, why, and to what the individuals respond.

The best-known example of social thermoregulation is undoubtedly that of honeybees. The data so far best fit the model that each bee regulates her own temperature, responding to what she feels, not what the others feel. As she responds to her own local thermal environment, however, she affects that of others around her. Her effect on them could be either helpful or harmful. But at this time in evolution the individual responses have been shaped so that those responses helpful to the colony have been accentuated. Thus, at high temperatures the bees cool themselves by water evaporation, but they do it *inside* the hive rather than outside, and the resulting thirsty bees then stimulate water gatherers. Additionally, fanning bees align themselves along pre-existing air streams to cool themselves convectively, but by doing it in the hive or at the hive entrance they aid the colony. At low temperatures, on the other hand, the bees regulate a high thoracic temperature (an individual response) in the presence of brood (a social advantage). Another example is the "comfort reaction" that calms bees at the periphery of a cluster so that they do not shiver unnecessarily (they can heat themselves if need be by crawling in). This response allows these bees to lower their thoracic temperature, thus saving its store of energy and therefore conserving colony energy stores. That this response is a product of natural selection is indicated by the lower expression of it in the African bees, which are more "nervous" than the European bees (see next section) and hence are better adapted for moving the colony and less well adapted for maintaining a winter cluster.

Geographical Distribution

In insects thermoregulation may be even less of a barrier to their distribution than it is in many other animals, such as birds or mammals, because insects may escape thermally unfavorable times if the inactive stage or stages (eggs, larvae, pupae) of their life cycle coincide with those times. Potential problems arise in those social species, like the honeybee, whose life cycle lacks a totally inactive stage.

Winter climate is the bottleneck that determines where honeybees exist, and for honeybees overwintering depends on thermoregulation, impressive as it is. For example, the European honeybee can prosper even in the Arctic in the summer, but it cannot survive there in winter. Some species of *Apis,* such as the Himalayan honeybee *Apis laboriosa,* use migration as an option for dealing with seasonal temperature extremes. In the United States a problem of practical interest is the northward expansion of the Africanized honeybee. Spreading from Brazil since the 1950s, these bees have already come through Central America and are now, in the mid-1990s, being reported in Texas. How far north will they come?

Individual bees of the African race of *Apis mellifera* thermoregulate as well and fly at ambient temperatures as low as those in which the European race is active. However, the European stock is better at *colony* thermoregulation for low-temperature survival, suggesting that temperature may limit the African bees' northward spread.

The African bees are adapted for rapid movement to exploit changing food resources. When food resources dwindle, the whole colony leaves its old hive and seeks a new area and a new hive. The bees are not very "fussy" in choosing a new nest site, probably because in their evolutionary past they have never needed to be selective. European bees, adapted to a northern environment, are less likely to find better conditions by moving elsewhere when the local bloom fails, as in the fall. Furthermore, their chances of survival depend much more strongly on also finding a nest site that is thoroughly evaluated before it is accepted. Thus, it is safer for European honeybees to stay than to take a chance on moving and finding another suitable nest site. Although the African bees' restless temperament is an adaptation to the environment, it likely also directly affects their capacity for colony thermoregulation.

The key is clustering. Bees in cluster reduce their metabolic rate by an order of magnitude below the rate when they are alone. When in a tight cluster, the metabolic heat generated by the individuals is conserved, and they no longer need to shiver at most ecologically relevant ambient temperatures. Indeed, metabolic rate may drop to resting levels. Clustered bees, possibly because they may warm up at any time simply by crawling into the group's interior, no longer defend a high body temperature and so conserve metabolic energy by coming to rest.

Both African and European bees experience a large, weight-specific reduction in metabolic rate when they are in groups, but as shown by the late Edward E. Southwick and colleagues, the reduction is much greater in the European than in the Africanized bees. The African bees continue to show "nervousness" and unwillingness to cluster tightly. That is, the higher metabolic rate of African honeybees in *groups* is due to poorer clustering. Indeed, Africanized bees *leave* their clusters, and their hive, even at air temperatures low enough for them to be quickly immobilized by the cold, and they may even fly out and thus be lost to the colony. The larger the winter colony, and the more cohesively it clusters, the better it can survive.

As a consequence of their high activity level, which compromises clustering behavior (but improves their industriousness in honey gathering), the Africanized bee will likely remain limited to a southerly distribution in the United States.

14

Insects in Man-Made Habitats

On a foggy morning once I watched a *Bombus vagans* worker weighing only 85 mg foraging on fireweed. She had presumably just crawled up a long dark tunnel from her underground nest and then emerged into the light. The morning chill—it was only 8°—would have made it difficult for her to fly, but she shivered until her thoracic temperature reached 32°C, and then she flew off to forage.

Some of the fireweed blossoms had recently been visited by other bees, and from them she got neither nectar nor pollen. But a few flowers had not yet been visited, and in them she found up to 1.85 mg sugar dissolved in nectar. She drained them and removed their pollen in a few quick motions, seldom requiring more than five seconds at each blossom.

I had placed a drop of viscous sugar syrup into some of the same fireweed flowers, and by random chance she found one of them. It would have taken her over two minutes to lap all of that syrup out of just one blossom, but I timed her with a stopwatch and after only one minute caught her and measured her thoracic temperature by inserting a small thermometer into her flight muscles. I knew from cooling experiments with a dead bee of the same size that while not producing heat she should have cooled to near ambient temperature. However,

A bumblebee foraging from fireweed, *Epilobium angustifolium.* When ambient temperature is low, it shivers while perching in the intervals between flights to individual flowers.

after one minute, this bee's thoracic temperature did not decline—instead, it had *increased* 2°C, to 34°C.

The conclusion was obvious. The bee *regulated* her thoracic temperature by maintaining a very high rate of heat production to oppose passive cooling after stopping flight. (Later studies in the laboratory showed that she did this by shivering with her flight muscles, and that thoracic temperature, whether high or low, depends in large part on her expectation of future food rewards.) Since I knew the temperatures (thoracic and ambient), the heat capacity of tissue (0.8 cal/g/°C), and the cooling rate as a function of the difference between her thoracic and the ambient temperature (2.1°C/min/°C difference), I was able to calculate the amount of heat expended by the bee in shivering to maintain 32°C (or to *any* thoracic temperature). This value was not a trivial investment for the bee, since this cost for maximal shivering rate turned out to be close to 0.04 watts, the same power output needed for flight.

By sampling the volumes and sugar concentrations of the fireweed flowers, I could also calculate how much energy the bee got back for her investment. The hotter (up to a point) she became, the faster she could forage and the more honey she could make. The calculations showed, for example, that the energy she invested in foraging on a chilly dawn was worthwhile because—provided competition at the flowers was low and provided she kept hot enough to handle at least ten flowers per minute—her profits would exceed costs several-fold. To her, time is honey, and body temperature is the means of translating one to the other.

With measurements like these I gained quantitative insights into the gains and losses of foraging behavior and the economy of the hive. My numbers could be used to estimate the food rewards that plants (or their neighbors) must provide to attract foragers for pollination services. In other words, the results applied to pollination ecology.

Determining the source and the amount of energy that a bumblebee uses to keep herself heated up, as calculated from her body temperature, is the sort of pursuit that has been held up to ridicule by politicians and journalists seeking to make a point about government spending. Indeed, in the late 1970s I was awarded $20,000 to study just this esoteric problem, and the project drew a headline in the *National Enquirer* for presumably wasting the taxpayers' money. About that time, students at Stanford University staged a "trivia contest," awarding first prize to a study of "the rectal temperature of the bumblebee." To me, of course, nothing could have been more exciting. I wrote *Bumblebee Economics* on the basis of this sort of trivia, and others have built on the perspective taken in this work. It became the inspiration, according to Michael Rothschild, a business consultant, for his highly acclaimed book *Bionomics*. That book resulted in the formation of the San Francisco–based Bionomics Institute, which promotes bio-logical thinking and holds annual conferences and now influences government and industry.

But to me the most rewarding application of bumblebee behavior, of which pollination energetics and the dynamics of coevolution is only one part, concerns the use of bees in crop production. Koppert Biological Systems of Rotterdam in the Netherlands is now the world leader in this area. The idea started with a Belgian veterinarian, Roland de Jonghe, whose hobby is bumblebee taxonomy. He was breeding bumblebees to try to differentiate sibling species (that is, those that look superficially similar). To distinguish species that could not be differentiated by eye, he used cross-breeding as a tool. A friend, a director at an agricultural school, who knew he raised bumblebees told him how tomato flowers are artificially pollinated in greenhouses by a labor-intensive process requiring an electric vibrator. To pollinate with a vibrator, a worker touches each plant with flowers individually to shake their pollen onto the flowers' pistils. Why not use bumble-bees instead, he asked. It was tried, and the results were instantly successful. De Jonghe started his company, Biobest. Within a year or so Biobest captured almost the entire tomato crop market in Holland. Competitors soon emerged, and one of these, Koppert, now produces even more bumblebees than Biobest.

Rearing bumblebees is not easy. There are tricks of the trade, and the secrets are zealously guarded. But thanks to my friend and colleague Hayo Velthius, whom I visited at the University of Utrecht, I was to get a tour of the facilities at Koppert Biological Systems.

The elegantly dressed secretary in the factory office building placed a call to the adjacent building to announce our arrival. Dr. Velthius and I then walked across the courtyard to enter the bumblebee factory. As we entered, we were met by Adriaan van Doorn, the slight, bearded man who has been the Director of Research and Development since the bumblebee operation started in 1987 and who is a former student of Velthius. I looked wide-eyed at the rows of gigantic stainless-steel tanks holding sugar syrup. As if reading my mind, which reeled at the enormity of all of this food energy, Dr. van Doorn said, "0.04 watts per bee!" That's not much, but there were a lot of bees,and this was a multi-million-dollar enterprise. In the large buildings stood endless racks of growing bumblebee colonies. Air temperature was strictly controlled so that the bees would tend brood, not just spend their time sitting on their brood to incubate it (p. 173) and burning up sugar supplies to boot.

Bumblebees are used primarily in greenhouse pollination of tomatoes (and, to a lesser extent, for some varieties of peppers, melons, strawberries, currants, and blueberries). Greenhouse production greatly increases the length of the growing season, and the use of bumblebees allows for pollination when wild pollinators are not available. Bumblebees are superior to other bees. They work in enclosed spaces and do not just fly to the light (the windows), as honeybees do. They also work at lower temperatures because of their superior thermoregulatory physiology.

Prior to the introduction of bumblebees as pollinators in 1987, greenhouse growers had to manually vibrate every plant twice a week. It was a tedious, boring job, and it could not duplicate natural pollination.

Bees respond to signals from the flowers themselves and pollinate them at the physiologically right time. Naturally pollinated flowers produce more aromatic fruit. When bee-pollinated tomatoes first came onto the market, they tasted better and were cheaper to produce. In short, they were so vastly superior that other growers had to follow suit, and within three years 95 percent of the tomato growers in Holland had switched to bumblebees. Now the use of bumblebees is nearly universal. In the winter of 1995, I found bumblebee-pollinated tomatoes from Holland, Israel, and Spain in the local Vermont supermarkets. They are bright red and aromatic, like home-grown tomatoes at last.

As an added bonus, the development of natural pollination runs parallel with the goal of a clean, pesticide-free environment, for without it the bumblebees cannot thrive. Thus, the bumblebee-pollinated greenhouses needed to be managed with biological rather than chemical control agents, and anyone buying bee-pollinated fruit from the greenhouse is also buying pesticide-free fruit.

In 1988 the venture with bumblebees "took off." Koppert alone now produces 120,000 colonies of bumblebees for sale per year, and it provides pollination and biological control services to growers in countries all over Europe, North Africa, Japan, and Korea. Subsidiary companies have also been set up in Israel and Canada. I suspect the United States will not lag far behind in the future.

In contrast to bees, many if not most insects that live with humans are unwelcome guests. The list is long: cockroaches, grain and flour beetles, fleas, Formosan and dry wood termites, clothes moths, book lice, silverfish, Argentine ants . . . All live close to or directly in their food, and most are of tropical origin and do not need to thermoregulate, nor have they evolved the capacity to do so. Unlike their hardy temperate-zone relatives, they live and reproduce only in narrowly circumscribed temperatures.

Our means of ridding ourselves of these pests has for decades included a potent chemical arsenal. We have built industries that are supported by this ongoing war. Unfortunately, however, many of the same chemicals that kill insects can kill us as well, or at least rob us of some life, for at the cellular level we are all practically identical, having evolved from the same root. Additionally, the chemical weapons don't always act where and when they are applied. For example, the July 1995 issue of *Environmental Health Perspectives* reports that although chlordane was banned in 1987, in some buildings in Houston, Texas, "significant indoor contamination persist[ed] through at least 1991." A good weapon is one that (like the bees' heat balls described in Chapter 12) can be applied very specifically and then removed, and it should work by capitalizing on a weakness or a

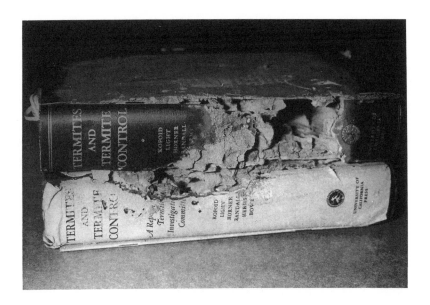

Thermal treatments might have saved these books (photographed in 1975) in library of the University of California at Berkeley.

difference. That seems a tall order. But there is, in fact, such a weapon. It's called thermal or temperature treatments.

In 1975 I tracked down an old book on termite control in the Berkeley library—only to find that it had been destroyed by the pests! Clearly, we humans were in trouble. Broadcasting pesticides in a public library to battle these book-eating invaders is not an attractive option: they don't really penetrate into thick dry books, but they readily reach the stratosphere and help deplete the ozone layer. But in 1977 a solution was found: temporarily deep-freezing the books. The solution of freezing to kill hidden insect pests was applied in Yale University's Beinecke Rare Book and Manuscript Library. The library's climate was maintained at environmental conditions (of 21°C ± 3°C and 50 percent relative humidity) ideal for humans and books—and, unfortunately, for insects. Book-eating insects include beetles, book lice, silverfish, and sometimes termites and cockroaches. Especially favored by some of these creatures are rare books, such as those with thick leather bindings. I now use the same deep-freeze method to get rid of insects destroying my valuable dried insect collections; I simply stick a specimen box in a freezer for a day. It saves months of smelling toxic, carcinogenic phenol crystals. That was the old way of "controlling" pests, and it seemed immutable. But most thermal applications are not, in the commercial economic sense, quite as simple as sticking a box into the freezer.

Many insects consume grain products as readily as we do, and perhaps even more readily than termites consume dry books. Chief among these consumers is the flour beetle, *Tribolium confusum*. It erupts in epidemic proportions wherever there is flour handling, large-scale dough production, dry mixing of cereals and cake mixes, bread or pizza making, and wherever grain and flour is stored. Methyl bromide, a potent poison, had always been the method of choice to control these pests in our food. However, the Pillsbury Company has recently done extensive tests with a control method called heat sterilization (HS) using temperatures of 54–60°C. This method proved to be extremely successful, and in 1992 the company established a program to use HS in all its flour-based operations. By 1994 it had eliminated the use of methyl bromide. Several other food companies (Nabisco, Quaker Oats, Con Agra, General Mills) have followed suit. Although thermal tolerance could gradually evolve in insect pests, it is very likely to be a much slower evolution than resistance to pesticides. Temperature affects *all* proteins in *all* living organisms, but pesticides can sometimes be detoxified by a single enzyme (an enzyme is a protein that catalyzes biochemical reactions). It is easier, evolutionarily, to alter just one enzyme than thousands of proteins all at once.

Insect pests don't just eat what is *in* our houses. Termites go all the way; they'll eat the house itself, or at least that part of it made of wood. But here again we humans are learning to turn the insects' own weapons against them.

In the mid-1980s the late Charles Forbes, a geologist at California State University at Domingues Hills, stopped a student of his, Joe Tallon, in the halls to ask: "What will liquid nitrogen do to bugs?" Tallon is a second-generation owner of an urban pest-control company, and the question spurred him to do an independent study. He first tested killing with cold by using liquid nitrogen on fleas in rodent burrows. It didn't work. Then he made a mock house for realistic tests on termites. And lo and behold, it worked! He could kill termites in situ! Simple enough. But can the method be applied *economically,* on a scale large enough to be practical?

Tallon then went to the high priest of urban entomology, Emeritus Professor Walter Ebeling at UCLA, and said to him: "Look at *this.* I've got something neat here!" Ebeling agreed and said: "History is being made . . . Here there may be an alternative to pesticides."

Heat treatments would later prove to be an even better method than cold treatments because pest insects are especially intolerant to heat. Besides, heat can be more easily applied. Ebeling and Tallon then patented different heat application methods, and history indeed *was* made.

The new methods that did not rely on pesticides were immediately warmly embraced by Tallon's customers. If successful, these methods could eventually result in the total phaseout of the use of pesticides in homes, a huge market. Mysteriously, however, about the same time, the California Department of Food and Agriculture refused to license the new methods, saying that they didn't work, and even declaring liquid nitrogen (for freeze-killing) an unregistered pesticide! (Close to 79 percent of the air we breathe is nitrogen.) Legal fumigants, such as methyl bromide and sulfuryl fluoride, surely do work—they have actually killed burglars unlucky enough to break into houses being treated. Tallon's company fought back, paying over a million dollars in legal fees trying to gain official approval for what amounts to cold air.

It was at this point that I got an excited phone call from Joe Tallon, who had just read my book, *The Hot-Blooded Insects*. He said that the first sentence of the prologue of that book ("No aspect of the physical environment is more important to insects than temperature") gave him hope. Maybe the book was even hefty enough to have some clout in court.

As a former resident of California I could attest to the problem of pests. We had cats and dogs, and our house near Berkeley was soon infested with fleas. Ultimately, we resorted to fumigating the house with pesticides to kill all of the uninvited guests. No matter that it was done by a registered Pest Control Vendor, we still worried about traces of toxic residues ending up in us. So we didn't do this often. But the fleas became unbearable, especially for our dog, Foonman, who developed allergies and skin rashes.

Then there were termites. Before buying a house in California, the bank financing it insists on a "termite inspection." Termites are an ever-present problem, and although exact figures are difficult to find, it is estimated that more than 200,000 chemical treatments are conducted annually in California for dry wood termites (*Incisitermes minor*). Now there are also the Formosan termites, which live sealed inside wood where toxic gases are slow to penetrate, if they ever do.

In the fall the paper wasps attach their nests under the eaves of houses and under doorways, and in California the pesticide trucks trafficked in and out of town in the pattern of morning milk deliveries of old. Only these trucks did not contain milk. Instead, they contained toxic brews. We let the wasp colonies die out naturally each fall, as we knew they would, but the neighbors didn't wait, because they thought the nests would keep growing larger and larger. Enter now Joe Tallon, an experienced and respected Pest Control Agent, who claims he is not anti-pesticide but who says: "We don't *need* to be toxic warriors. If the client chooses, we can be *thermal* warriors. It's good for business."

The chemical industry was not pleased with this pronouncement. It wields influence in the Structural Pest Control Board, which dictates what works and what control is to be allowed and what would not be allowed. Ostensibly the Board would allow only what was safe and what would work, which, of course, were the chemicals pushed by the big chemical manufacturers. "Fighting pests without chemical treatments, with temperature," Tallon told me, "is according to them a ridiculous idea. It's just too simple to be effective." But his study with Forbes and Ebeling had already showed that, to the contrary, thermal treatment is very effective.

Presumably to settle the argument, university researchers then got state money to see if temperature treatments do indeed work. They compared the effectiveness of chemical fumigation, microwaves, and thermal treatments for controlling dry wood termites (*I. minor*). Laboratory experiments had been done before, but in these new tests, costing $300,000, the researchers sought to simulate field conditions by building a mock structure, the Villa Termiti. The twenty-by-twenty-foot structure had doors and windows on all four sides, an exterior of stucco walls, and a shingled roof. Douglas fir boards infested with known numbers of termites could be inserted into the Villa, and removed after treatment, so that the effectiveness of the treatments could be assessed.

Given the difficulty of objectively comparing one method against another, the researchers chose vendors for their particular specialty in controlling dry wood termites. After each treatment the damage to the installed termites was assessed. For the heat treatment forty-eight termite-infested boards were positioned prior to the heat treatment, twelve in the attic, twenty-four in the dry-wall area behind walls, and twelve under the house. The Villa was then wrapped in a tarp, which contained holes for ventilation, and four propane heaters were used to raise the internal temperature of the wood in the Villa (monitored with thermocouples) to 48°C (120°F). It took five hours to reach this temperature, and it was maintained for one hour. Then the test boards were removed and inspected for dead termites. In total, only 39 termites out of 3,000 used for the test were still alive, and after four weeks only 14 of these 39 were still living. It has previously been shown that mild heat treatment kills the termites' gut symbionts required to digest cellulose. Therefore, the immediate mortality is less than the final result because the remaining live termites may ultimately have starved to death.

Most people might suppose that this was a victory for thermal treatments, but it wasn't announced that way. As may happen in any study, the way it is interpreted can depend on perspective or bias. In this case, CNN reported that although thermal treatments work, pesticides do best because they killed more of the termites (at least outright). Thus, the pesticide battle continues, and pesticides are still used. Now the new rubric is Integrated Pest Management (IPM). It *sounds* like a reasonable compromise. You use heat treatments. Then you apply the same chemicals to "clean up." Everyone wins, especially pesticide salesmen.

It is difficult to argue with the success of the thermal strategy that Tallon patented in 1987 and 1988. You can't improve upon it, in theory, only in your methods of application. It will likely not be long before an arsenal of heat-application methods will become available to zap the colony of hornets on your front porch, the fleas in your rug, and maybe the termites in your walls. We'll be the thermal warriors, using the very methods to control insects that some of them use on *their* enemies. Then we'll not only have learned about insects, we'll have learned *from* them.

As we should. After all, they already know what works—they patented it millions of years ago. It's not harmful to us, and it's environmentally correct.

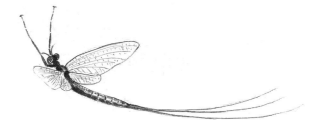

Epilogue

Activity in all living things is highly constrained by temperature, and most animals have means of avoiding extremes of hot and cold. They do so in part by stabilizing (regulating) body temperature in the face of environmental fluctuations. In other words, they attempt to achieve a free and independent life, unconstrained by perturbations in ambient temperature.

Insects are different from us in many ways, but most tangibly in that they sense a thermal world much different from our own. They experience much greater extremes in temperature than we do. Because of their small size, some insects could heat some 10°C while perching in a sunfleck—and cool just as quickly in the shade. They lack the thermal inertia of our large bulk, and so they face much greater problems than we do in thermoregulation. Those that do thermoregulate despite their small size may therefore be considered to have greater thermoregulatory control than larger animals have.

My purpose in writing this book was to survey what insects can do and how they do it, and why. Performance is constrained by limits, and insects have evolved their biochemical machinery so that they can fly at tissue temperatures from 0°C to nearly 50°C, which is near the limits presently dictated by biochemistry. Different insect species have evolved, without regard to phylogenetic origin, to regulate body temperature within a range that is related to body mass and thermal environment. Indeed, an insect's mass dictates much of the thermal environment its tissues experience. The central lesson to be drawn from the fact that insects as a whole regularly fly at a *range* of muscle temperatures is that their muscles have been biochemically tailored through evolution to operate at temperatures near what they have experienced, and that insects now attempt to achieve, by behavior and physiology, those temperatures to which they have been adapted. Undoubtedly, the same principles apply to vertebrates.

Body temperatures are affected by an extraordinarily large array of variables, and insects manipulate a good many of these variables. Their complex life cycles allow them to control the season and habitat in which they are active. They also exploit diurnal and spatial thermal changes to their advantage. But predators are often just a step behind, necessitating behavioral and physiological adjustments.

Body temperature at any one given thermal environment is regulated by adjustments of mechanisms for both heat production and heat loss. The insects' unique body plan has allowed the evolution of mechanisms found nowhere else in the animal world. However, remarkably close analogies exist in vertebrate animals.

For most insects, thermoregulation (as opposed to biochemical adaptation to ambient temperature) is a consequence of the evolution of ever more powerful flight. The heat produced by powerful flight muscles resulted in a high level of endothermy, and thermoregulation evolved when insects developed the capacity to warm up their muscles prior to flight. These same endothermic insects then also evolved to fly over a *range* of air temperatures, principally by developing warm-up and heat-loss mechanisms. These latter developments were largely a consequence of large mass in conjunction with the high metabolic demands of flight. Because of the unavoidable heating in flight, these insects had to adapt their biochemical flight machinery to operate at the high end of the temperatures experienced during flight. And these thoracic temperatures then had to be regulated. Most small insects, on the other hand, had no need for thermoregulation in flight because they did not experience the high muscle temperatures reached in larger insects.

Warm-up prior to either long or intermittent flight is accomplished in many diurnal insects by a variety of postural adjustments that maximize surface area available to solar heating and simultaneously minimize convective cooling. Shivering warm-up occurs in most large insects, whether they are nocturnal or diurnal. It involves only the thoracic (wing) muscles, and thoracic temperature rises steeply in any one warm-up bout as the flight muscles are activated by neural signals that vary only slightly from those used in flight itself.

These central themes provide a core around which an almost bewildering variety of ecologically relevant strategies and counterstrategies have evolved. The thermoregulatory capacities of any one insect species affects its interactions with others, and vice versa. As a result, we find the evolution of thermal warriors—insects whose ability to control body temperature is a weapon in their competition for survival. Some of these insects engage in thermal "arms races" with their competitors, and with their predators and prey. Some of these arms races have been waged, and won, against the presumed champions of thermoregulation, the homeothermic birds. Some insects, on the other hand, would better be described as "thermal nurturers." The thermoregulatory abilities of insects that live in social groups, such as bees, have evolved to enhance and enlarge the capacity for cooperation among individuals; their use of temperature control for the benefit of colony mates is one reason these social insects have evolved to become superorganisms with perhaps unprecedented social organization.

It had until recently been thought that insects were simple, cold-blooded poikilotherms, despite often tantalizing hints to the contrary. We now recognize them as thermal warriors of surprising sophistication. It is not their crudeness that was blinding, but ours.

Selected Readings

1. From Cold Crawlers to Hot Flyers

Brogniart, C. 1894. Recherches pour servir á l'histoire des insectes fossiles des temps primaires. *Thèses de la Faculté des Sciences de Paris,* no. 821, pp. 1–494.

Carpenter, R. M. 1966. The Lower Permian insects of Kansas. Part II. The orders Protothoptera and Orthoptera. *Psyche* 73:46–88.

Dumont, J. P. C., and R. M. Robertson. 1986. Neural circuits: An evolutionary perspective. *Science* 233:849–853.

Heinrich, B. 1977. Why have some animals evolved to regulate a high body temperature? *American Naturalist* 111:623–640.

Heinrich, B., and T. M. Casey. 1978. Heat transfer in dragonflies: "Fliers" and "perchers." *Journal of Experimental Biology* 74:17–36.

Holdsworth, R. 1941. The wing development of *Pteronarcys proteus* Newman. *Journal of Morphology* 70:431–461.

Kluger, M. J. 1979. *Fever: Its Biology, Evolution and Function.* Princeton, N.J.: Princeton University Press.

Kohshima, S. 1984. A novel cold-tolerant insect found in a Himalayan glacier. *Nature* 30:225–227.

Kukalova, J. 1968. Permian mayfly nymphs. *Psyche* 75:310–327.

Kukalova-Peck, J. 1978. Origin and evolution of insect wings and their relation to metamorphosis, as documented by the fossil record. *Journal of Morphology* 156:53–126.

————— 1985. Ephemeroid wing venation based upon new gigantic Carboniferous mayflies and basic morphology, phylogeny, and metamorphosis of pterygote insects (Insecta, Ephemerida). *Canadian Journal of Zoology* 63:933–955.

————— 1987. New Carboniferous Diplura, Monura, and Thysanura, the hexapod ground plan, and the role of thoracic side lobes in the origin of wings (Insecta). *Canadian Journal of Zoology* 65:2327–2345.

Marden, J. H., and M. G. Kramer. 1994. Surface-skimming stoneflies: A possible intermediate stage in insect flight evolution. *Science* 266:427–430.

May, M. 1982. Heat exchange and endothermy in Protodonata. *Evolution* 36:1051–1058.

Robertson, R. M., K. G. Pearson, and H. Reichert. 1982. Flight interneurons in the locust and the origin of insect wings. *Science* 217:177–179.

Ruben, J. 1995. The evolution of endothermy in mammals and birds: From physiology to fossils. *Annual Review of Physiology* 57:69–95.

Shear, W. A., and J. Kukalova-Peck. 1990. The ecology of Paleozoic terrestrial arthropods: The fossil evidence. *Canadian Journal of Zoology* 68:1807–1834.

Somero, G. N. 1995. Proteins and temperature. *Annual Review of Physiology* 57:43–68.

Stavenga, D. G., P. B. W. Schwering, and J. Tinbergen. 1993. A three-compartment model describing temperature changes in tethered flying blowflies. *Journal of Experimental Biology* 185:325–333.

Wigglesworth, V. B. 1976. The evolution of insect flight. In *Insect Flight,* ed. R. C. Rainey. *Royal Entomological Society of London, Symposia* 7:255–269.

Wootton, R. J. 1981. Paleozoic insects. *Annual Review of Entomology* 26:319–344.

2. Heat Balance

Casey, T. M., M. L. May, and K. R. Morgan. 1985. Flight energetics of euglossine bees in relation to morphology and wing stroke frequency. *Journal of Experimental Biology* 116:271–289.

Heinrich, B. 1974. Thermoregulation in endothermic insects. *Science* 185:747–756.

Heinrich, B., and T. M. Casey. 1973. Metabolic rate and endothermy in sphinx moths. *Journal of Comparative Physiology* 82:195–206.

Kammer, A. E., and B. Heinrich. 1978. Insect flight metabolism. *Advances in Insect Physiology* 13:133–228.

May, M. 1982. Heat exchange and endothermy in Protodonata. *Evolution* 36:1051–1058.

————— 1995. Dependence of flight behavior and heat production on air

temperature in the green darner dragonfly *Anax junius* (Odonata: Aeshnidae). *Journal of Experimental Biology* 198:2385–2392.

Roubik, D. W. 1993. Tropical pollinators in the canopy and understory: Field
data and theory for stratum "preferences." *Journal of Insect Behavior*
6:659–673.

Stone, G. N. 1993. Thermoregulation in four species of tropical solitary bees:
The role of size, sex and altitude. *Journal of Comparative Physiology*
B163:317–326.

3. The Flight Motor

Boettiger, E. G. 1960. Insect flight muscles and their basic physiology. *Annual
Review of Entomology* 5:1–15.

Casey, T. M. 1981. Energetics and thermoregulation of *Malacosoma americanum* (Lepidoptera: Lasiocampidae) during hovering flight. *Physiological Zoology* 54:362–371.

Cooper, A. J. 1993. Limitations of bumblebee flight performance. Dissertation.
University of Cambridge, England.

Ellington, C. P., K. E. Machin, and T. M. Casey. 1990. Oxygen consumption
of bumblebees in forward flight. *Nature* 347:472–473.

Esch, H. 1988. The effects of temperature on flight muscle potentials in
honeybees and cuculiinid winter moths. *Journal of Experimental Biology*
135P:109–117.

Harrison, J. M. 1986. Caste-specific changes in honeybee flight capacity.
Physiological Zoology 59:175–187.

Heglund, N. C., M. A. Fedak, C. R. Taylor, and G. A. Cavagna. 1982. Energetics
and mechanics of terrestrial locomotion. IV. Total mechanical energy
changes as a function of speed and body size in birds and mammals.
Journal of Experimental Biology 97:57–66.

Heinrich, B., and T. P. Mommsen. 1985. Flight of winter moths near 0°C.
Science 228:177–179.

Josephson, R. K., and R. D. Stevenson. 1991. The efficiency of a flight muscle
from the locust *Schistocerca americana*. *Journal of Physiology*
442:413–429.

Marden, J. H. 1989. Bodybuilding dragonflies: Costs and benefits of maximizing flight muscle. *Physiological Zoology* 62:505–521.

———— 1995. Evolutionary adaptation of contractile performance in muscles
of moths that fly with widely divergent body temperatures. *Journal of
Experimental Biology* 198:2087–2094.

Pringle, J. W. S. 1954. The mechanism of the myogenic rhythm of certain
insect fibrillar muscles. *Journal of Physiology* 124:269–291.

Stevenson, R. D., and R. K. Josephson. 1990. Effects of operating frequency and temperature on mechanical power output from moth flight muscle. *Journal of Experimental Biology* 149:61–79.

Watt, W. B. 1977. Adaptation of specific loci. I. Natural selection on phospho-glucose isomerase of *Colias* butterflies: Biochemical and population aspects. *Genetics* 87:177–194.

———— 1983. Adaptation at specific loci. II. Demographic and biochemical elements in the maintenance of the *Colias* PG1 polymorphism. *Genetics* 103:691–725.

4. Warm-Up by Shivering

Casey, T. M., J. R. Hegel, and C. S. Buser. 1981. Physiology and energetics of pre-flight warm-up in the Eastern tent caterpillar moth *Malacosoma americanum*. *Journal of Experimental Biology* 94:119–135.

Dorsett, D. A. 1962. Preparation for flight by hawk-moths. *Journal of Experimental Biology* 39:579–588.

Esch, H., and J. Bastian. 1968. Mechanical and electrical activity in the indirect flight muscles of the honey bee. *Zeitschrift für Vergleichende Physiologie* 58:429–440.

Esch, H., and F. Goller. 1991. Neural control of honeybee fibrillar muscles during shivering and flight. *Journal of Experimental Biology* 159:419–431.

Esch, H., C. Goller, and B. Heinrich. 1991. How do bees shiver? *Naturwissenschaften* 78:325–328.

Heinrich, B. 1972. Temperature regulation in the bumblebee, *Bombus vagans:* A field study. *Science* 175:185–187.

Heinrich, B., and G. A. Bartholomew. 1971. An analysis of pre-flight warm-up in the sphinx moth, *Manduca sexta*. *Journal of Experimental Biology* 55:223–239.

Heinrich, B., and A. E. Kammer. 1973. Activation of the fibrillar muscles in the bumblebee during warm-up, stabilization of thoracic temperature and flight. *Journal of Experimental Biology* 58:677–688.

Heinrich, B., and C. Pantle. 1975. Thermoregulation in small flies (*Syrphus* sp.): Basking and shivering. *Journal of Experimental Biology* 62:599–610.

Heinrich, B., and M. J. Heinrich. 1983. Size and caste in temperature regulation by bumblebees. *Physiological Zoology* 56:552–562.

Ikeda, K., and E. G. Boettiger. 1965. Studies on the flight mechanism in insects. II. The innervation and electrical activity of the fibrillar muscles of the bumblebees, *Bombus*. *Journal of Insect Physiology* 11:779–789.

Kammer, A. E. 1967. Muscle activity during flight and warm-up in Lepidoptera. *Journal of Experimental Biology* 47:277–295.

——— 1968. Motor patterns during flight and warm-up in Lepidoptera. *Journal of Experimental Biology* 48:89–109.

——— 1970. A comparative study of motor patterns during pre-flight warm-up in hawkmoths. *Zeitschrift für Vergleichende Physiologie* 70:45–56.

——— 1970. Thoracic temperature, shivering, and flight in the monarch butterfly, *Danaus plexippus*. *Zeitschrift für Vergleichende Physiologie* 68:334–344.

Kammer, A. E., and B. Heinrich 1972. Neural control of bumblebee fibrillar muscle during shivering. *Journal of Comparative Physiology* 78:337–345.

——— 1974. Metabolic rates related to muscle activity in bumblebees. *Journal of Experimental Biology* 61:219–227.

——— 1978. Insect flight metabolism. *Advances in Insect Physiology* 13:133–228

Krogh, A., and E. Zeuthen. 1941. The mechanism of flight preparation in some insects. *Journal of Experimental Biology* 18:1–10.

Leston, D., J. W. S. Pringle, and D. C. S. White. 1965. Muscular activity during preparation for flight in a beetle. *Journal of Experimental Biology* 42:409–414.

Morgan, K. R., and B. Heinrich. 1987. Temperature regulation in bee- and wasp-mimicking syrphid flies. *Journal of Experimental Biology* 133:59–71.

Neville, A. C., and T. Weis-Fogh. 1963. The effect of temperature on locust flight muscle. *Journal of Experimental Biology* 40:123–136.

Srygley, R. B. 1994. Shivering and its cost during reproductive behaviour in Neotropical owl butterflies, *Caligo* and *Opsiphanes* (Nympholidae: Brassolinae). *Animal Behavior* 47:23–32.

5. Warm-Up by Basking

Dreisig, H. 1995. Thermoregulation and flight activity in territorial male graylings, *Hipparchia semele* (Satyridae), and large skippers, *Ochlodes renata* (Hesperidae). *Oecologia* 101:167–176.

Fraenkel, D. G. 1930. Die Orientierung von *Schistocerca gregaria* zu strahlender Wärme. *Zeitschrift für Vergleichende Physiologie* 13:300–313.

Heinrich, B. 1972. Thoracic temperatures of butterflies in the field near the equator. *Comparative Biochemistry and Physiology* 43A:459–467.

——— 1975. Thermoregulation in bumblebees: II. Energetics of warm-up and free flight. *Journal of Comparative Physiology* 96:155–166.

——— 1986. Thermoregulation and flight activity of the satyr, *Coenonympha ornata* (Lepidoptera: Satyridae). *Ecology* 67:593–597.

——— 1986. Comparative thermoregulation of four montane butterflies of different mass. *Physiological Zoology* 59:616–626.

——— 1990. Is "reflectance" basking real? *Journal of Experimental Biology* 154:31–43.

Kevan, P. G., and J. D. Shorthouse. 1970. Behavioral thermoregulation by High Arctic butterflies. *Arctic* 23:268–279.

May, M. 1976. Thermoregulation and adaptation to temperature in dragonflies (Odonata: Anisoptera). *Ecological Monographs* 46:1–32.

May, M. 1990. Thermal adaptations of dragonflies revisited. *Advances in Adonatology* 5:71–88.

——— 1995. Simultaneous control of head and thoracic temperature by the green darner dragonfly *Anax junius* (Odonata: Aeshnidae). *Journal of Experimental Biology* 198:2373–2384.

Morgan, K. R., T. S. Shelly, and L. S. Kimsey. 1985. Body temperature regulation, energy metabolism, and foraging in light-seeking and shade seeking robber flies. *Journal of Comparative Physiology* B155:561–570.

Polcyn, D. M., and M. A. Chappell. 1986. Analysis of heat transfer in *Vanessa* butterflies: Effects of wing position and orientation to wind and light. *Physiological Zoology* 59:706–716.

Ravenscroft, N. O. M. 1994. Environmental influences on mate location in male chequered skipper butterflies, *Carterocephalus polaemon* (Lepidoptera: Hesperidae). *Animal Behavior* 47:1179–1187.

Remmert, H. 1985. Crickets in sunshine. *Oecologia* (Berlin) 68:29–33.

Schmitz, H., and L. T. Wasserthal. 1993. Antennal thermoreceptors and wing-thermosensitivity of heliotherm butterflies: Their possible role in thermoregulatory behavior. *Journal of Insect Physiology* 39:1007–1019.

Tracy, C. R., B. J. Tracy, and D. Dobkin. 1979. The role of posturing in behavioral thermoregulation by black dragons (*Hagenius selys;* Odonata). *Physiological Zoology* 52:565–571.

Vielmetter, W. 1954. Die Temperaturregulation des Kaisermantels in der Sonnenstrahlung. *Naturwissenschaften* 41:535–536.

Vogt, D. F., and B. Heinrich. 1983. Thoracic temperature variations in the onset of flight in dragonflies (Odonata: Anisoptera). *Physiological Zoology* 56:236–241.

Wasserthal, L. T. 1975. The role of butterfly wings in regulation of body temperature. *Journal of Insect Physiology* 21:1921–1930.

——— 1983. Haemolymph flows in the wings of pierid butterflies visualized by vital staining (Insecta, Lepidoptera). *Zoomorphology* 103:177–192.

Watt, W. B. 1968. Adaptive significance of pigment polymorphism in *Colias* butterflies. I. Variation of melanin pigment in relation to thermoregulation. *Evolution* 22:437–458.

Winn, A. F. 1916. Heliotropism in butterflies; or turning toward the sun. *Canadian Entomologist* 48:6–9.

6. Cooling Off

Baird, J. M. 1986. A field study of thermoregulation in the carpenter bee *Xylocopa virginica virginica* (Hymenoptera: Anthophoridae). *Physiological Zoology* 59:157–167.

Bartholomew, G. A., J. R. B. Lighton, and G. N. Louw. 1985. Energetics of locomotion and patterns of respiration in tenebrionid beetles from the Namib Desert. *Journal of Comparative Physiology* B155:155–162.

Bartholomew, G. A., and J. R. B. Lighton. 1986. Endothermy and energy metabolism of a giant tropical fly, *Pantopthalmus tabaninus* Thunberg. *Journal of Comparative Physiology* B156:461–467.

Chappell, M. A. 1982. Temperature regulation of carpenter bee *(Xylocopa californica)* foraging in the Colorado Desert of southern California. *Physiological Zoology* 55:267–280.

Coelho, J. R., and A. J. Ross. 1996. Body temperature and thermoregulation in two species of yellowjackets, *Vespula germanica* and *V. maculifrons*. *Journal of Comparative Physiology* B (in press).

Cooper, P., W. M. Schaffer, and S. L. Buchmann. 1985. Temperature regulation of honey bees *(Apis mellifera)* foraging in the Sonoran Desert. *Journal of Experimental Biology* 114:115.

Edney, E. B., and R. Barrass. 1962. The body temperature of the tse-tse fly, *Glossina morsitans* Westwood (Diptera, Muscidae). *Journal of Insect Physiology* 8:469–481.

Ellington, C. P. 1984. The aerodynamics of hovering insect flight. VI. Lift and power requirements. *Royal Society of London, Philosophical Transactions* B305:145–181.

Hadley, N. F., E. C. Toolson, and M. C. Quinlan. 1989. Regional differences in cuticular permeability in the desert cicada *Diceroprocta apache:* Implications for evaporative cooling. *Journal of Experimental Biology* 141:219–230.

Hegel, J. I., and T. M. Casey. 1982. Thermoregulation and control of head temperature in the sphinx moth, *Manduca sexta. Journal of Experimental Biology* 101:1–15.

Heinrich, B. 1970. Thoracic temperature stabilization by blood circulation in a free-flying moth. *Science* 168:58–582.

——— 1970. Nervous control of the heart during thoracic temperature regulation in a sphinx moth. *Science* 169:606–607.

——— 1971. Temperature regulation of the sphinx moth, *Manduca sexta.* II. *Journal of Experimental Biology* 54:141–166.

——— 1976. Heat exchange in relation to blood flow between thorax and abdomen in bumblebees. *Journal of Experimental Biology* 54:561–585.

——— 1980. Mechanisms of body-temperature regulation in honeybees, *Apis mellifera*. II. Regulation of thoracic temperature at high air temperatures. *Journal of Experimental Biology* 85:73–87.

Heinrich, B., and T. M. Casey. 1978. Heat transfer in dragonflies: "Fliers" and "perchers." *Journal of Experimental Biology* 74:17–36.

Heinrich, B., and S. L. Buchmann. 1986. Thermoregulatory physiology of the carpenter bee, *Xylocopa varipuncta*. *Journal of Comparative Physiology* B156:557–562.

Lindauer, M. 1954. Temperaturregulierung und Wasserhaushalt im Bienen-staat. *Zeitschrift für Vergleichende Physiologie* 36:391–432.

Morgan, K. R. 1985. Body temperature regulation and terrestrial activity in the ectothermic beetle *Cicindela tranquebarica*. *Physiological Zoology* 58:29–37.

Morgan, K. R., T. S. Shelly, and L. S. Kimsey. 1985. Body temperature regulation, energy metabolism, and foraging in light-seeking and shade-seeking robber flies. *Journal of Comparative Physiology* B155:561–570.

Nicolson, S. W., and G. D. Louw. 1982. Simultaneous measurement of evaporative water loss, oxygen consumption and thoracic temperature during flight in a carpenter bee. *Journal of Experimental Biology* 222:287–296.

Prange, D. H. 1990. Temperature regulation by respiratory evaporation in grasshoppers. *Journal of Experimental Biology* 154:463–474.

Toolson, E. C. 1987. Water prolifigacy as an adaptation to hot deserts: Water loss rates and evaporative cooling in the Sonoran Desert cicada, *Diceroprocta apache* (Homoptera: Cicadidae). *Physiological Zoology* 60:379–385.

Toolson, E. C., P. D. Ashby, R. W. Howard, and D. Stanley-Samuelson. 1994. Eicosanoids mediate control of thermoregulatory sweating in the cicada, *Tibicen dealbatus* (Insecta: Homoptera). *Journal of Comparative Physiology* B. 164:278–285.

7. Form and Function

Bartholomew, G. A., and B. Heinrich. 1973. A field study of flight temperatures in moths in relation to body weight and wing load. *Journal of Experimental Biology* 58:123–135.

Casey, T. M., and J. R. Hagel. 1981. Caterpillar setae: Insulation for an ectotherm. *Science* 214:1131–1133.

Casey, T. M., and B. A. Joos. 1983. Morphometrics, conductance, thoracic temperature, and flight energetics of noctuid and geometrid moths. *Physiological Zoology* 56:160–173.

Church, N. S. 1960. Heat loss and body temperature of flying insects. II. Heat conduction within the body and its loss by radiation and convection. *Journal of Experimental Biology* 37:186–212.

Dreisig, H. 1981. The rate of predation and its temperature dependence in a tiger beetle, *Cicindela hybrida*. *Oikos* 36:196–202.

———— 1990. Thermoregulatory stilting in tiger beetles, *Cicindela hybrida* L. *Journal of Arid Environments* 19:297–302.

Freudenstein, K. 1928. Das Herz und das Zirkulationssystem der Honigbiene (*Apis mellifica* L.). *Zeitschrift für Wissenschaftliche Zoologie* 132:404–475.

Hadley, N. F., T. D. Schultz, and A. Savill. 1988. Spectral reflectances of three tiger beetle subspecies (*Neocicindela perhispida*): Correlation with habitat substrate. *New Zealand Journal of Zoology* 15:343–346.

Heinrich, B. 1976. Heat exchange in relation to blood flow between thorax and abdomen in bumblebees. *Journal of Experimental Biology* 54:561–585.

———— 1979. *Bumblebee Economics*. Cambridge, Mass.: Harvard University Press.

———— 1980. Mechanisms of body-temperature regulation in honeybees, *Apis mellifera*. II. Regulation of thoracic temperature at high air temperatures. *Journal of Experimental Biology* 85:73–87.

———— 1987. Thermoregulation by winter-flying endothermic moths. *Journal of Experimental Biology* 127:313–332.

Holm, E., and E. B. Edney. 1973. Daily activity of Namib Desert arthropods in relation to climate. *Ecology* 54:45–56.

Kukal, O., B. Heinrich, and J. G. Duman. 1988. Behavioural thermoregulation in the freeze-tolerant Arctic caterpillar *Gynaephora groenlandica*. *Journal of Experimental Biology* 138:181–193.

May, M. L., and T. M. Casey. 1983. Thermoregulation and heat exchange in euglossine bees. *Physiological Zoology* 56:541–551.

Morgan, K. R. 1985. Body temperature regulation and terrestrial activity in the ectothermic beetle *Cicindela tranquebarica*. *Physiological Zoology* 58:29–37.

Nicholson, S. W., G. A. Bartholomew, and M. K. Seeley. 1994. Ecological correlates of locomotion speed, morphometrics and body temperature in three Namib Desert tenebrinid beetles. *South African Journal of Zoology* 19:131–134.

Parker, M. A. 1982. Thermoregulation by diurnal movement in the barberpole grasshopper (*Dactylotum bicolor*). *American Midland Naturalist* 107:228–237.

Vogt, D. F., and B. Heinrich. 1994. Abdominal temperature regulation by Arctic bumblebees. *Physiological Zoology* 66:257–269.

Watt, W. B. 1968. Adaptive significance of pigment polymorphism in *Colias* butterflies. I. Variation of melanin pigment in relation to thermoregulation. *Evolution* 22:437–458.

Whitman, D. W. 1987. Thermoregulation and daily activity patterns in a black desert grasshopper, *Taeniopoda eques. Animal Behavior* 35:1814–1826.

Wille, A. 1958. A comparative study of the dorsal vessel of bees. *Annals of the Entomological Society of America* 51:538–546.

8. Conserving Energy

Heinrich, B., and M. J. Heinrich. 1983. Heterothermia in foraging workers and drones of the bumblebee, *Bombus terricola. Physiological Zoology* 56:563–567.

Heinrich, B., and E. McClain. 1986. "Laziness" and hypothermia as a foraging strategy in flower scarabs (Coleoptera: Scarabaeidae). *Physiological Zoology* 59:273–282.

Masters, A. R., S. B. Malcolm, and L. P. Brower. 1988. Monarch butterfly *(Danaus plexippus)* thermoregulatory behavior and adaptations for over-wintering in Mexico. *Ecology* 69:458–467.

Schmaranzer, S., and A. Stabentheimer. 1988. Variability of the thermal behavior of honeybees on a feeding place. *Journal of Comparative Physiology* B15:135–141.

Sivinski, J. 1978. Intrasexual aggression in the stick insects, *Diapheromera veliei* and *D. covilleae,* and sexual dimorphism in the Phasmatodea. *Psyche* 85:395–406.

Thornhill, R. 1976. Reproductive behavior of the lovebug, *Plecia nearctica* (Diptera: Bibionidae). *Entomological Society of America, Annals* 69:843–845.

Waddington, K. D. 1990. Foraging profits and thoracic temperature of honey bees *(Apis mellifera). Journal of Comparative Physiology* B160:325–329.

Woods, W. A., Jr., and R. D. Stevenson. 1996. Time and energy costs of mating for the sphinx moth, *Manduca sexta. Journal of Experimental Biology* (in press).

9. Why Do Insects Thermoregulate?

Carruthers, R. I., T. S. Larkin, H. Firstencel, and Z. Feng. 1992. Influence of thermal ecology on the mycosis of rangeland grasshoppers. *Ecology* 73:190–204.

Dreisig, H. 1995. Thermoregulation and flight activity in territorial male graylings, *Hipparchia semele* (Satyridae), and large skippers, *Ochlodes renata* (Hesperidae). *Oecologia* 191:167–176.

Gronenberg, W., J. Tautz and B. Hölldobler. 1993. Fast trap jaws and giant neurons in the ant *Odontomachus*. *Science* 262:561–563.

Heinrich, B. 1977. Why have some animals evolved to regulate a high body temperature? *American Naturalist* 111:623–640.

———— 1986. Thermoregulation and flight activity of the satyr, *Coenonympha ornata* (Lepidoptera: Satyridae). *Ecology* 67:593–597.

———— 1987. Thermoregulation by winter-flying endothermic moths. *Journal of Experimental Biology* 127:313–332.

Heinrich, B., and T. P. Mommsen. 1985. Flight of moths near 0°C. *Science* 228:177–179.

Heinrich, B., and D. F. Vogt. 1993. Thermoregulation in Arctic bumble-bees: Regulation of abdominal temperature. *Physiological Zoology* 66:257–269.

Kukal, O., B. Heinrich, and J. G. Duman. 1988. Behavioural thermoregulation in the freeze-tolerant Arctic caterpillar *Gynaephora groenlandica*. *Journal of Experimental Biology* 138:181–193.

Morgan, K. R., and T. E. Shelly. 1988. Body temperature regulation in desert robber flies (Diptera: Asilidae). –*Ecological Entomology* 14:419–428.

Nicolson, S. W. 1987. Absence of endothermy in flightless dung beetles from southern Africa. *South African Journal of Zoology* 22:23–324.

Oertli, J. J., and M. Oertli. 1990. Energetics and thermoregulation of *Popillia japonica* Newmand (Scarabaeidae, Coleoptera) during flight and rest. *Physiological Zoology* 63:921–937.

Shine, R. 1994. Young lizards can be bearable. *Natural History,* January, pp. 34–38.

Tinbergen, N., B. J. D. Meeuse, L. K. Boerema, and W. W. Varossieau. 1942. Die Balz des Samtfalters, *Eumenis (Satyrus) semele* (L). *Zeitschrift für Tierpsychologie* 5:182–226.

Vogt, D. F., and B. Heinrich. 1994. Abdominal temperature regulation by Arctic bumblebees. *Physiological Zoology* 66:257–269.

Ybarrando, B. A. 1995. Habitat selection and thermal preferences in two species of water scavenging beetles (Coleopter: Hydrophilidae). *Physiological Zoology* 68:749–771.

10. Strategies for Survival

Barnes, B. M., J. L Barger, J. Seares, P. C. Tacguard, and G. L. Zuercher. 1996. Overwintering physiology and behavior in yellow-jacket queens *(Vespula vulgaris)* and green stink bugs *(Meadorus lateralis)* in subarctic Alaska. *Physiological Zoology* (in press).

Bartholomew, G. A., and B. Heinrich. 1978. Endothermy in African dung beetles during flight, ball making, and ball rolling. *Journal of Experimental Biology* 73:65–83.

Elder, H. Y. 1971. High frequency muscles used in sound production by a katydid. II. Ultrastructure of the singing muscles. *Biological Bulletin* (Woods Hole) 141:434–448.

Heath, J. E., and P. J. Wilkin. 1970. Body temperature and singing in the katydid, *Neoconocephalus robustus* (Orthoptera, Tettigoniidae). *Biological Bulletin* (Woods Hole) 138:272–285.

Heinrich, B. 1972. Energetics of temperature regulation and foraging in a bumblebee, *Bombus terricola* Kirby. *Journal of Comparative Physiology* 77:49–64.

Heinrich, B., and G. A. Bartholomew. 1979. Roles of endothermy and size in inter- and intraspecific competition for elephant dung in an African dung beetle, *Scarabaeus laevistriatus*. *Physiological Zoology* 52:484–496.

Heinrich, B., and C. Pantle. 1975. Thermoregulation in small flies (*Syrphus* sp.): Basking and shivering. *Journal of Experimental Biology* 62:595–610.

Heller, K. 1986. Warm-up and stridulation in the bushcricket, *Hexacentrus unicolor* Serville (Orthoptera, Conocephalidae, Listroscelidinae). *Journal of Experimental Biology* 126:97–109.

Josephson, R. K. 1973. Contraction kinetics of the fast muscles used in singing by a katydid. *Journal of Experimental Biology* 59:781–801.

———— 1985. The mechanical power output of a tettigoniid wing muscle during singing and flight. *Journal of Experimental Biology* 117:357–368.

Josephson, R. K., and H. Y. Elder. 1968. Rapidly contracting muscles used in sound production by a katydid. *Biological Bulletin* (Woods Hole) 135:409.

Josephson, R. K., and D. Young. 1979. Body temperature and singing in the bladder cicada. *Cystosoma saundersii*. *Journal of Experimental Biology* 80:69–81.

———— 1985. A synchronous muscle with an operating frequency greater than 500 Hz. *Journal of Experimental Biology* 118:185–208.

Marden, J. H. 1995. Evolutionary adaptation of contractile performance in muscle of ectothermic winter-flying moths. *Journal of Experimental Biology* 198:2087–2094.

Srygley, R. B. 1994. Shivering and its cost during reproductive behaviour in Neotropical owl butterflies, *Caligo* and *Upsiphanes* (Nymphalidae: Brassolinae). *Animal Behaviour* 47:23–32.

Stevens, E. D., and R. K. Josephson. 1977. Metabolic rate and body temperature in singing katydids. *Physiological Zoology* 50:31–42.

Ybarrondo, B. A., and B. Heinrich. 1994. Thermoregulation and response to competition in the African dung ball-rolling beetle *Kheper nigroaeneus* (Coleoptera: Scarabaeidae). *Physiological Zoology* 69:35–48.

11. Thermal Arms Races

Chai, P., and R. B. Srygley. 1989. Predation and the flight, morphology, and temperature of neotropical rainforest butterflies. *American Naturalist* 135:748–765.

Chappell, M. A. 1983. Thermal limitations to escape responses in desert grasshoppers. *Animal Behavior* 31:1088–1093.

Esch, H. 1988. The effects of temperature on flight muscle potentials in honeybees and cuculiinid winter moths. *Journal of Experimental Biology* 135:109–117.

Goller, F., and H. Esch. 1990. Comparative study of chill-coma temperatures and muscle potentials in insect flight muscles. *Journal of Experimental Biology* 150:221–231.

Harkness, R., and R. Wehner. 1977. *Cataglyphis. Endeavor* 1:115–121.

Heath, J. E., and R. K. Josephson. 1970. Body temperature and singing in the katydid, *Neoconocephalus robustus* (Orthoptera, Tettigoniidae). *Biological Bulletin* (Woods Hole) 138:272–285.

Heinrich, B. 1979. Thermoregulation of African and European honeybees during foraging, attack, and hive exits and returns. *Journal of Experimental Biology* 80:217–229.

——— 1984. Strategies of thermoregulation and foraging in two vespid wasps, *Dolichovespula maculata* and *Vespula vulgaris. Journal of Comparative Physiology* B154:175–180.

——— 1987. Thermoregulation by winter-flying endothermic moths. *Journal of Experimental Biology* 127:313–332.

Holm, E., and E. B. Edney. 1973. Daily activity of Namib Desert arthropods in relation to climate. *Ecology* 54:45–56.

Kenagy, G. J., and R. D. Stevenson. 1982. Role of body temperature in the seasonality of daily activity in tenebrionid beetles of eastern Washington. *Ecology* 63:1491–1503.

Lighton, J. R. B., and R. Wehner. 1993. Ventilation and respiratory metabolism in the thermophilic desert ant, *Cataglyphis bicolor* (Hymenoptera, Formicidae). *Journal of Comparative Physiology* B163:11–17.

Marden, J. H. 1995. Large-scale changes in thermal sensitivity of flight performance during adult maturation in a dragonfly. *Journal of Experimental Biology* 198:2095–2102.

Marden, J. H., and P. Chai. 1991. Aerial predation and butterfly design: How palatability, mimicry, and the need for evasive flight concerns mass allocation. *American Naturalist* 138:15–36.

Marsh, A. C. 1985. Thermal responses and temperature tolerance in a diurnal desert ant, *Ocymyrmex barbiger. Physiological Zoology* 58:629–636.

Morgan, K. R. 1987. Temperature regulation, energy metabolism, and mate-searching in rain beetles (*Plecoma* spp.): Winter-active, endothermic scarabs (Coleoptera). *Journal of Experimental Biology* 128:107–122.

Schweitzer, D. 1974. Notes on the biology and distribution of Cuculiinae (Noctuidae). *Lepidopterists' Society, Journal* 28:5–21.

Srygley, R. B., and P. Chai. 1990. Predation and the elevation of thoracic temperature in brightly-colored, neotropical butterflies. *American Naturalist* 135:766–787.

Syrlykke, A. and A. E. Treat. 1995. Hearing in winter moths. *Naturwissenschaften* 82:382–384.

Toolson, E. C. 1987. Water profligacy as an adaptation to hot deserts: Water loss rates and evaporative cooling in the Sonoran Desert cicada, *Diceroprocta apache* (Homoptera: Cicadidae). *Journal of Experimental Biology* 131:439–444.

Toolson, E. C., and N. E. Hadley. 1987. Energy-dependent facilitation of transcuticular water flux contributes to evaporative cooling in the Sonoran Desert cicada, *Diceroprocta apache* (Homoptera: Cicadidae). *Journal of Experimental Biology* 131:439–444.

Wehner, R., A. C. Marsh, and S. Wehner. 1992. Desert ants on a thermal tightrope. *Nature* 357:586–587.

12. Heat Treatments

Boorstein, S. M., and P. W. Ewald. 1987. Cost and benefits of behavioral fever in *Melanoplus sanguinipes* infected by *Nosema acridophagus*. *Physiological Zoology* 60:586–595.

Bronstein, M. S., and W. E. Conner. 1984. Endotoxin-induced behavioral fever in the Madagascar cockroach, *Gramphadorhina protentosa*. *Journal of Insect Physiology* 30:327–330.

Carruthers, R. I., T. S. Larkin, H. Firstencel, and Z. Feng. 1992. Influence of thermal ecology on the mycosis of rangeland grasshoppers. *Ecology* 73:190–204.

Eisner, T., and J. Meinwald. 1966. Defensive secretions of arthropods. *Science* 153:1341–1350.

Feder, J. H., J. M. Rossi, J. Solomon, N. Solomon, and S. Lindquist. 1992. The consequences of expressing Hsp70 in *Drosophila* cells at normal temperatures. *Genes & Development* 6:1402–1413.

Feder, M. E. In press. Ecological and evolutionary physiology of stress proteins and the stress response: The *Drosophila melanogaster* model. In *Phenotypic and Evolutionary Adaptations to Temperature,* ed. I. A. Johnston and A. F. Bennett. Cambridge: Cambridge University Press.

Gehring, W. J., and R. Wehner. 1995. Heat shock protein synthesis and thermotolerance in *Cataglyphis,* an ant from the Sahara Desert. *Proceedings of the National Academy of Sciences USA* 92:2994–2998.

Heinrich, B. 1979. Thermoregulation in African and European honeybees during foraging, attack, and hive exits and returns. *Journal of Experimental Biology* 80:217–229.

Kluger, M. J. 1979. *Fever: Its Biology, Evolution and Function.* Princeton, N.J.: Princeton University Press.

Lindquist, S. 1986. The heat-shock response. *Annual Review of Biochemistry* 55:1151–1191.

Louis, C., M. Jordan, and M. Cabanac. 1986. Behavioral fever and therapy in a rickettsia-infected orthopteran. *American Journal of Physiology* 250:R991–R995.

Martin, J., M. Mayhew, T. Langer, and F. U. Hartle. 1993. The reaction cycle of GroEL and GroES in chaperonin-assisted protein folding. *Nature* 366:228–233.

McClain, E., P. Magnuson, and S. J. Warner. 1988. Behavioral fever in a Namib Desert tenebrionid beetle, *Onymacris plana. Journal of Insect Physiology* 34:279–284.

Müller, C. B., and P. Schmid-Hempel. 1993. Exploitation of cold temperature as defense against parasitoids in bumblebees. *Nature* 363:65–67.

Ono, M., I. Okada, and M. Sasaki. 1987. Heat production by balling in the Japanese honeybee, *Apis cerana japonica,* as a defensive behavior against the hornet, *Vespa simillima xanthoptera* (Hymenoptera: Vespidae). *Experimentia* 43:1031–1032.

Ono, M., T. Igarashi, E. Ohno, and M. Sasaki. 1995. Unusual thermal defense by a honeybee against mass attack by hornets. *Nature* 377:334–336.

Ritossa, F. M. 1962. A new puffing pattern induced by temperature shock and DNP in *Drosophila. Experimentia* 18:571–573.

Velazquez, J. M., and S. Lindquist. 1984. Hsp70: Nuclear concentration during environmental stress and storage during recovery. *Cell* 36:655–662.

Welte, M. A., J. M. Tetrault, R. P. Dellavalle, and S. L. Lindquist. 1993. A new method for manipulating transgenes: Engineering heat tolerance in a complex, multicellular organism. *Current Biology* 3:842–853.

13. Heating and Cooling the Nest

Brian, M. V. 1973. Temperature choice and its relevance to brood survival and caste determination in the ant *Myrmica rubra* L. *Physiological Zoology* 46:245–252.

Bruman, F. 1928. Die Luftzirkulation im Bienenstock. *Zeitschrift für Vergleichende Physiologie* 8:366–370.

Coenen-Stass, D., B. Schaarschmidt, and I. Lamprecht. 1980. Temperature distribution and colorimetric determination of heat production in the nest of the wood ant, *Formica polyctena* (Hymenoptera, Formicidae). *Ecology* 61:238–244.

Corkins, C. L. 1932. The temperature relationship of the honeybee cluster under controlled temperature conditions. *Journal of Economic Entomology* 25:820–825.

Elton, C. 1932. Orientation of the nests of *Formica turncorum* in north Norway. *Journal of Animal Ecology* 1:192–193.

Franks, N. R. 1989. Thermoregulation in army ant bivouacs. *Physiological Zoology* 14:397–404.

Free, J. B., and H. Y. Spencer-Booth. 1960. Chill-coma and cold death temperatures of *Apis mellifica*. *Entomologia Experimentalis et Applicata* 3:222–230.

Gay, F. J., and J. H. Calaby. 1970. Termites from the Australian region. In *Biology of Termites*, vol. 2, ed. K. Krishna and F. M. Weesner. New York: Academic.

Heinrich, B. 1981. The mechanisms and energetics of honeybee swarm temperature regulation. *Journal of Experimental Biology* 91:25–55.

Loos, R. 1964. A sensitive anemometer and its use for the measurement of air currents in the nests of *Macrotermes natalensis* (Haviland). In *Études sur les Termites Africains*, ed. A. Bouillon. Paris: Masson.

Lüscher, M. 1961. Air-conditioned termite nests. *Scientific American* 205:138–145.

Michener, C. D., R. B. Lange, J. J. Bigarella, and R. Salamuni. 1959. Factors influencing the distribution of bees' nests in earth banks. *Ecology* 29:207–217.

Moritz, R. F. A., and P. Kryger. 1994. Self-organization of circadian rhythms in groups of honeybees (*Apis mellifera* L.). *Behavioral Ecology and Sociobiology* 34:211–215.

Ofer, J. 1970. *Polyrachis simplex*, the weaver ant of Israel. *Insectes Sociaux* 17:49–82.

Rosengren, R., W. Fortelius, K. Lindström, and A. Luther. 1987. Phenology and causation of nest heating and thermoregulation in red wood ants of the *Formica rufa* group studied in coniferous forest habitats in southern Finland. *Annales Zoologici Fennici* 24:147–155.

Seeley, T. D. 1977. Measurements of nest cavity volume by the honeybee *(Apis mellifera)*. *Behavioral Ecology and Sociobiology* 2:201–227.

Seeley, T. D., and B. Heinrich. 1981. Regulation in the nests of social insects. In *Insect Thermoregulation*, ed. B. Heinrich. New York: Wiley.

Southwick, E. E., D. W. Roubik, and J. M. Williams. 1990. Comparative energy balance in groups of Africanized and European honey bees: Ecological implications. *Comparative Biochemistry and Physiology* 97A:1–7.

Tschinkel, W. R. 1987. Seasonal life history and nest architecture of a winter-active ant, *Prenolepis imparis. Insectes Sociaux* 34:143–164.

Underwood, B. A. 1990. Seasonal nesting cycle and migration patterns of the Himalayan honey bee *Apis laboriosa. National Geographic Research* 6:276–290.

Vogt, D. F. 1986. Thermoregulation in bumblebee colonies. I. Thermoregulatory versus brood-maintenance behaviors during acute changes in ambient temperature. *Physiological Zoology* 59:55–59.

Waloff, N., and R. E. Blacklith. 1962. The growth and distribution of the mounds of *Lasius flavus* (Fabricius) (Hym. Formicidae) in Silwood Park, Berkshire. *Journal of Animal Ecology* 31:421–437.

Weyrauch, W. 1936. Das Verhalten sozialer Wespen bei Nestüberhitzung. *Zeitschrift für Vergleichende Physiologie* 23:51–63.

14. Insects in Man-Made Habitats

Heinrich, B. 1972. Temperature regulation in bumblebees, *Bombus vagans:* A field study. *Science* 175:185–187.

Forbes, C., and W. Ebeling. 1986. Update: Liquid nitrogen controls dry wood termites. *IPM Practitioner* 8:1–4.

——— 1987. Update: Use of heat for the elimination of structural pests. *IPM Practitioner* 9:1–6.

Nesheim, K. 1984. The Yale non-toxic method of eradicating book-eating insects by deep-freezing. *Restaurator* 6:147–164.

Rust, M. K., and B. J. Cabrera. 1995. The effect of temperature and humidity on the movement of the western dry wood termite. Unpublished manuscript, Department of Entomology, University of California, Riverside.

Index

Schmitz, Helmut, 53
Scramble competition, 129
Setae, 82–84
Sexual selection, 123–137
Shivering, 35–46, 124; functions of, 43–44; energetics of, 44–47, 136. *See* Warm-up (shivering)
Singing, 131, 152–153
Southwick, Edward E., 183
Srygley, Robert B., 136
Stevenson, Robert D., 32, 143
Stilting, 64–87
Supercooling, 125–126
Superorganisms, 181, 199
Surlykke, Anne M., 145

Tallon, Joseph C., 193, 194–196
Temperature: measurement of, 17, 187; tolerance of, 141–143, 160–161
Thermal preferences, 116–117, 141–142
Thermal treatments (and tolerances), 191–196
Thermophilic insects, 134, 140–141
Thermoregulation, 2, 106; effect of body mass on, 107; evolution of, 3, 17, 38–39, 69, 107, 116–117; and

growth, 110; and mating, 109, 134; and set-points, 2
Tinbergen, Niko, 109
Toolson, Eric, 75
Torpor, 2
Treat, Asher E., 145

van Doorn, Adriaan, 188
Velthius, Hayo, 188
Vogt, F. Daniel, 115

Warm-up (shivering), 132, 137; behavior, 45–46; energetics of, 44–47, 186–187; evolution of, 38, 41; for heat production, 36; mechanisms of, 38–43; neural activation during, 40; rate of, 45–47
Wasserthal, Lutz, 53
Watt, Ward B., 34, 85
Wehner, Rüdiger, 141
Wheeler, Peter, 66–67
Wigglesworth, Sir Vincent, 5
Wing-loading, 80, 120
Wings: evolution of, 3–11; functions of, 4; venation of, 9–10

Ybarrondo, Brent A., 116, 130